THE NEW SCIENCE

A MAJOR PARADIGM SHIFT

THE NEW SCIENCE

A MAJOR PARADIGM SHIFT

VOLUME II

NOEL HUNTLEY

Copyright © 2019 by Noel Huntley.

ISBN: Softcover   978-1-7960-4028-9
          eBook   978-1-7960-4040-1

All rights reserved. No part of this book may be reproduced or transmitted in any form or by any means, electronic or mechanical, including photocopying, recording, or by any information storage and retrieval system, without permission in writing from the copyright owner.

Any people depicted in stock imagery provided by Getty Images are models, and such images are being used for illustrative purposes only. Certain stock imagery © Getty Images.

Print information available on the last page.

Rev. date: 06/12/2018

To order additional copies of this book, contact:
Xlibris
1-888-795-4274
www.Xlibris.com
Orders@Xlibris.com
798435

TABLE OF CONTENTS

PART ONE
    NEW SCIENCE FOUNDATIONS
1. Introduction and Overview   11
2. The Whole and the Part   18
3. The Limitations of Scientific Methodology and Objectivity   25
4. The Importance of Tests of Truth in Physics   37
5. Self-Referencing Systems and the Relative Zero   42
6. The Absolute   53
7. The Infinite Fractal   62
8. The Interface System   74

PART TWO
    MIND AND CREATION PHYSICS
9. Formatting Systems   81
10. Cybernetics and Creative and Automatic Awareness   89
11. Action Concepts and the Learning Pattern   99
12. Body, Mind and Consciousness   108
13. Creation Physics   117
14. Artificial and Unnatural Realities   128
15. The Healing Process   135

PART THREE
    SCIENCE IN THE ARTS AND SPORTS
16. Science and the Arts   149
17. The New Science in Sports and Athletics   175
18. The Nature of Intelligence   192

PART FOUR
> **SUPERSPACE**
>> 19. The Universal Vortex and Electromagnetism   217
>> 20. The Gravitational Field   237
>> 21. Inertia, a Covariant Aether and the Velocity of Light   250
>> 22. Epilogue   266

APPENDIX A   Integration, Differentiation and Quantum Regeneration   273
APPENDIX B   The Laser and Quantum Regeneration   280
APPENDIX C   Visualising the 4D Vortex   283
APPENDIX D   Quantum Reduction at different Orders of the Fractal Scale   286
APPENDIX E   Scalar Energies   290
APPENDIX F   Table: Orthodox and New Science Compared   299
APPENDIX G   Knowledge Structures   304

BIBLIOGRAPHY   319
INDEX   322
Advert   329

LIST OF ILLUSTRATIONS

FIGURE 1	The ladder analogy	13
FIGURE 2	Three fundamental forms of energy	23
FIGURE 3	Example of relative zero	46
FIGURE 4	Universal vortices	65
FIGURE 5(a)	Random particle fractals	70
FIGURE 5(b)	Fractal orders indicated	70
FIGURE 6(a)	Underlying fractal orders	72
FIGURE 6(b)	Inner fractals revealed externally	72
FIGURE 7	Subjectivity and objectivity	77
FIGURE 8	Feedback loop	92
FIGURE 9	Simple interface mechanism	93
FIGURE 10	The ultimate cybernetic feedback system	95
FIGURE 11	Interface between idea and mechanism	100
FIGURE 12	Computer bits and the learning pattern	101
FIGURE 13	Action concepts and computer bits	102
FIGURE 14	Ratio of consciousness to unconsciousness	110
FIGURE 15	Higher aspects of self	112
FIGURE 16	Linear and nonlinear processes (triad principle)	122
FIGURE 17	From Absolute to total objectivity	124
FIGURE 18	Training schedule	180
FIGURE 19	Simple demo of computer program	184
FIGURE 20	Vector model of aether	218
FIGURE 21	Nodes and wave intersections	221
FIGURE 22	Basic vortex	223

FIGURE 23	4D to 3D vortex	225
FIGURE 24	Electrons in magnetic field	229
FIGURE 25	Magnetic field around electron flow	230
FIGURE 26	Creation of electric field	232
FIGURE 27	3D to 4D space curvature	238
FIGURE 28	Repeat of 4D -3D vortex	238
FIGURE 29	Gravity—'attraction'	239
FIGURE 30	Gravitational vortex pair	241
FIGURE 31	Gravity, antigravity and spacetime	242
FIGURE 32	Gravity, doughnut configuration	243
FIGURE 33	Quantum regeneration and emergent software	272
FIGURE 34	Laser	279
FIGURE 35	Visualising 4D vortex	282
FIGURE 36	Secondary 4D vortex	283
FIGURE 37	Spot-light analogy for universe structure	287
FIGURE 38	Quantum reduction at different fractal orders	289
FIGURE 39	Electromagnetic and scalar standing waves	295

PART ONE

NEW SCIENCE FOUNDATIONS

1.

INTRODUCTION AND OVERVIEW

The real universe is infinitely more sophisticated, intelligent and complex than is revealed by mainstream science and education.

Let us first present a preview of what this book is about, which is a sequel to volume I, *The Emerging New Science*. There will unavoidably be a certain amount of re-presentation of data but this will aid clarification.

By 'New Science' we don't mean one separate from existing science, but more in the way of an expansion of orthodox knowledge to include all facets of life and existence, in particular, those sections excluded by our science. Again when we say 'expansion' this does not mean immediate enlightenment of all subjects but a view of the big picture; a paradigm that embraces all aspects of the universe and experience: an extended framework to cover the cosmos of universes, mind, consciousness, spirituality, religion, the paranormal, etc. Inevitably this will involve a major paradigm shift.

Thus the New Science offers a framework which will embrace all phenomena from the cosmic scale of manifestations. However, inevitably there emerges two extreme paths: 1) to either focus on consolidation of known data (integrating, coordination, organising, etc.), but neglecting expansion into more advanced contexts when these become apparent, or 2) to focus on the expansion cycle towards the ultimate context, leaving the details to be 'filled' in by science for, say, the next millions of years. The ideal is, of course, a balance between the two. We shall see that the current scientific approach is to over-focus on (1) and suffer the consequences of becoming stuck in what would otherwise be a proper evolution of knowledge.

Moreover, as far as possible we want to make this as accessible as possible to the general reader. The essential purpose of this book is to present a scientific theory of Creation. What do we mean by theory of Creation? Religion has, or has its basis in, creation theories, such as the biblical one about God creating the universe in six days, etc. The significant difference between science and religion is that whereas religion recognises that all manifestation comes from God, which means creation began with the highest order, that is, the maximum truth and thus it may be regarded as a top-down theory, science endeavours to prove that all began with a Big Bang or at least randomness, or chaos, or even nothing, the lowest level/order, and is referred to as a 'bottom-up' theory. Figure 1 gives a simple picture of this using the ladder analogy.

Much of this argument between science and religion exists in, and has answers within, the simple relationship between the whole and the part. The whole corresponds to the highest order, and the part, the lowest. Which came first? See Section 2

Thus the fundamental difference between science and religion is that religion is 'top-down' and science is a 'bottom-up'. This is the primary relationship and difference between science and religion. Later we shall show that both science and religion are not strictly natural subjects due to an artificial separation of the subjective and objective realities. Figure 1 shows simplistically and diagrammatically, using the ladder analogy, the relationship between the scope of science and religion, and the greater context of the New Science.

The main point we wish to make here is that in nature or natural creation the whole comes first. It is already inherent, such as a planet not being just made up of atoms but has underlying wholeness, a single quantum state. Nevertheless, although we are able to artificially create the whole from the parts by resonance and harmonics, this is quantum *regeneration,* not generation. The whole energy state is already inherent in what we might call a virtual subjective state, an underlying background fractal gradient going from maximum fragmentation at the bottom, to maximum wholeness at the top, the Absolute (see later section on the infinite fractal).

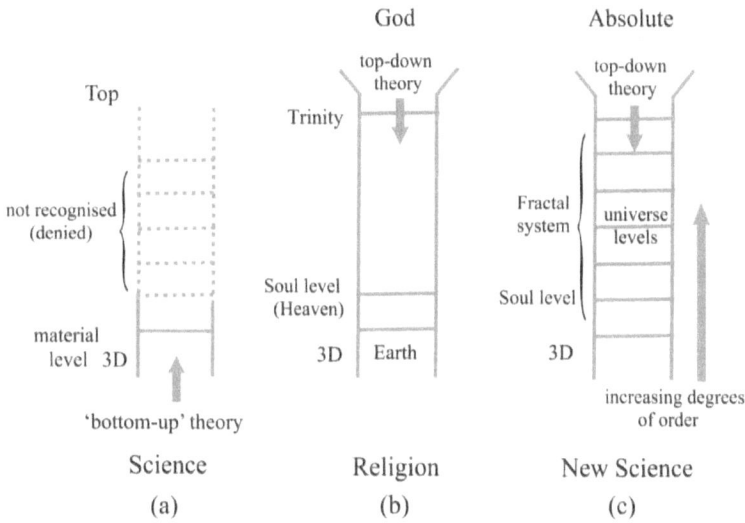

FIGURE 1: Ladder analogy: comparing orthodox science, religion and the New Science.

This quantum regeneration is a continuous and universal process and we shall encounter more explanations throughout the book. What else is telling us that the whole, or greatest whole, comes first before its parts? Fractals are probably the most important example. Fractals are normally expressed as self-similar forms on different scales. But this nowhere near reveals the true meaning and purpose of fractals. The discovery of fractals is one of the most important in science, but science does not explain the reasons for them; why everything obeys the law of fractals. The conventional definition does not include the all-important organisation within organisation or a scale of degrees of order, such as the fractal system of a tree, in which the degree of order increases as we go from twigs to branches to tree trunk. It is clear to anyone that the trunk of the tree comes before the branches or twigs (creation begins with the greatest order).[1]

Let's take another example of the action and joint arrangement in the human arm, which is quite similar fractally to the tree branch; that is, twigs corresponding to fingers. In learning skills, involving the whole arm, such as piano-playing technique, it can be observed and proved that causation passes down from shoulder (joint movement) to elbow, to wrist, to fingers. Causative action or control cannot pass up from fingers to shoulder. This is saying that a pattern of movements learned by the higher joint will transfer to a lower joint but not the reverse. That is, if the elbow learns a pattern it will transfer down to the wrist and fingers, but not transfer up to the shoulder. However, a lower fractal level can send information of its condition to a higher level, and the higher level will, if necessary, act on it and pass down information. Thus the shoulder level has all the information below it; and the elbow all information below it, etc. The higher the fractal level the more dominant it is.[2]

A similar fractal condition arises for universes and multi-universes. A lower level can't detect a higher level (from that level), but our minds and consciousness are also fractal and we can potentially access the higher levels of the universe by, in effect, moving up our focus in the fractal scale to our higher parts/aspects (see later section on the mind and consciousness).

Thus again we can see that in the case of the fractal arm the higher level comes first. The tree is a simple obvious example; the trunk comes before the twigs, or branches. The higher orders exist in their own right and are causative over the lower levels (parts, particles).

We shall show that the scientific method has inherent limitations, as clearly revealed by quantum physics some 80 years ago. It was shown that all measurements and observations have an inner, universal subjective component or, in other words, what is observed is in the context of whom or what is doing the observing. This means the experimental set-up has a limited context that forms the basis of the reference system upon which results of measurements are evaluated (everything that can be measured is contextual; in the context of something else (a higher order)—this is covered in Section 3).

Fortunately science has more powerful methods for testing truth; what we refer to as the tests of truth in physics (see Section 4). We shall see that these have a higher rating than the experimental method in revealing the degree of how relative or absolute is the result of a theory. We shall see that the tests of truth indicate the level of truth by evaluating the degree of the absolute factor; for example, if one theory poses that an object interacting in a medium is a separate construction from that medium, the theory has a degree of dualism. [The two bodies are not intrinsically related by the theory, such as in Einstein's general theory of relativity, compared with the ocean model (supported by quantum physics), in which 'shapes and variations', such as waves and eddies, represent mass; which eliminates the dualism characteristic.]

We shall see in the later section how these tests-of-truth features bring us closer to the Absolute reference point; also that this is the fixed zero and ultimate still point. It is the final step in a series of relative fractal levels of truth, directing us towards the Absolute. The cosmos is far more complex than education reveals.

Just as we underestimate the sophistication of the universe system we also fail to recognise how complex the human being is (Part 2). We shall show that man's mind and consciousness is engaged in a fractal system. Compared with science's simplistic brain-and-body model of the human, the true human is infinitely more elaborate.

If the reader is having difficulty believing in the reality of higher-fractal levels of existence, in particular, why isn't science detecting these 'solid' universes, let us remind the reader of the familiar human experience, that of our limited range of hearing compared with, say, dogs. Everyone knows that when the frequency of sound becomes sufficiently high we don't hear it, but dogs have extended hearing. Thus what sounds we don't hear, don't exist for us but can exist for other observers with a greater range (perception) of detection. Similarly this can apply to mass. We know from quantum physics (also metaphysics, New Age, etc.) that underlying the illusions of 'empty' space and solid objects are frequency patterns—that is, packets of vibrations, patterns of waves. Within a certain frequency range these wave configurations are visible and solid to humans; what we recognise as matter. If we imagine raising the frequencies (as with the above case of sound) these waves forming mass will go outside our range of detection; that is, will be invisible and one could even physically move through them. If we were to increase our frequencies to that level, the objects, structures, would become solid to us, and so on for higher levels.

Thus there is nothing strange about invisibility and intangibility being real when viewed within its appropriate context. We shall also show later how these limitations arise in the scientific experimental method.

As research and development occurs within the various disciplines of science it eventually becomes clear that physics is the basic science—even biology or genetics would be engaged in the interaction and functions of particles, waves, forces, space and time, and complex systems of these, manifesting in the holographic fractal patterns and orders within nature and the universe. Having established the ubiquitous role of physics, the question arises as to what is the basic subject within physics. It is simply the understanding of how energy works—or if one prefers: how mechanisms work (note this covers everything that is quantifiable and not the non-quantifiables, which we shall discuss later).

Notes
1. Article: *The Fractal Tree.* www.nhbeyondduality.org.uk.
2. Experimental psychologists did thorough experiments many years ago on this 'ranking' system of the learning level of the joints of the arm, and proved the validity of what is being stated here.

2.

THE WHOLE AND THE PART

Orthodox science teaches that the part comes before the whole; the New Science reveals that basically the whole precedes the part. A lower order cannot create a higher order.

As we study this subject in depth we may recognise that the whole/part relates to unity/separateness, quality/quantity, integration/differentiation, and subjective/objective. Some of the most basic underlying principles of what we can call creation physics are revealed from a study of the relationship between the whole and the part. Science posits that the part comes first, in contrast to the spiritual, religious and metaphysical view that the whole comes first. However, we must point out that in artificial systems, such as a motor vehicle, the whole equals the sum of the parts. In natural systems, such as atoms, molecules, cells, planets, stars, etc., we are claiming that the whole is always dominant over the parts. The whole may not always appear to come first, such as in the growth of living things, but they grow from a blue print, and even an atom or a planet will have a blueprint. A blueprint is a higher-order structure. Thus the whole came first. In general, where the part *appears* to come first, there was already a dormant virtual-state whole present (higher dimensionally and higher frequency).

Science has not recognised quantum regeneration, already mentioned, which is another method of manifesting the whole, quite magically from the parts (such as entraining the frequencies of the parts) about which we shall say more—it is one of the most important principles of creation. However, let us continue with this inquiry about the subject of whether, or when, the whole is greater than the sum of the parts and, in particular, whether the whole comes first.

Science teaches that there is no difference between the wholeness of natural structures and parts stuck together by forces. In the latter case of parts held together by forces we only have a composite unity; a simulated wholeness. There is in fact a potentially infinite difference between the true whole and the part. All the major conceptual issues and constructive difficulties in the pursuit of knowledge at the most basic level could be resolved by an understanding of the relationship between the whole and the part.

Science considers all wholes as having been built from parts, as in our artificial systems, such as a motor car. Surely an organism grows from the part to the whole. However, we know that the growth follows a blueprint, a template or DNA. This is a high order which guides the lower order of the parts to form a meaningful whole.

It is a logical impossibility for a lower order to create a higher order. The complex interdependence of parts within a system, such as the evolution of the eye, wouldn't by chance all interrelate fortuitously to complete the function of the eye—as in fact admitted by Darwin. Also an artificial product does not have a true natural whole, but a composite whole made up of parts stuck together by forces, such as a motor car in which the parts are held together by forces (welding or nuts and bolts). In this case, the whole begins with the parts and equals the sum of the parts. This limits the whole to 3D and readily is acceptable to the intellect that can now reduce the whole to parts, understand the parts, and then understand the whole. This does not apply to any natural entity. In all natural bodies, such as atoms, planets, stars, etc., and of course life in general, the whole is greater than the sum of the parts. The higher order is already inherent, just as are virtual particles, which come and go but can transform into real permanent particles if provided with enough energy. However, if the frequencies of particles or waves (such as in the laser; see Appendix B) are put into resonance, a higher order is quantum regenerated—of higher frequency (we shall come back to this).

We might point out that if an object has a function, such as an engine or even a nut and bolt, then conceptually the whole is greater than the sum of the parts. For instance, the engine can create motion and do work—the mere sum of the separate parts doesn't achieve this. However, we are only interested in the hardware energy/matter relationship in comparing the part and the whole and not in abstract functions (which are relative to a civilisation's specific development or preferences).

We are interested in the creation level, the hardware. We must go deeper; for instance, what happens to the energy and the mass? Fractal systems are an obvious example of the greater whole being primary to the part. For instance, the tree design is a simple and easy-to-understand fractal creation. It is obvious that the branch comes before the twig. The energy transduces down from the trunk region (the greatest whole) to the branches and finally twigs.

A very different example is revealed by the principles of coordination of, say, the human body. The author has studied this subject extensively (topic of doctorate in experimental psychology). Let's say an individual is learning to play the piano. Anyone is familiar with the requirement of achieving independence of finger action. But how many people would know that to achieve independence (degree of separateness) one must develop integration of movements. The more movement, or number of movements, that can be spanned by a learning pattern, that is, as a whole, the more differentiation capability there will be, that is, greater independence; thus the greater the integration, the greater the differentiation. Control over the part, differentiated from another part, cannot occur until the learning patterns enable the mind/consciousness to grasp more wholeness—more motion in space and time. One part cannot connect to another part in a *controlled* system if there is no integrated wholeness of the parts. There must be an underlying interconnectedness between parts if there is to be control over the parts (this is a nonlinear system). This is a holographic mechanism to which we shall return. Also, this applies to true collectives: the greater the wholeness, the greater is the potential power connected to the separate parts or individuals.

The learning pattern is a 4D holographic template that houses programs and converts nonlinear information into linear information. We don't need to explain all the functions of this here. The learning-pattern template holographically spans a certain amount of movements in space and time*—the mind computer bits can be coordinated giving a kinaesthetic sense in the muscles of the immediate future positions. This occurs quite automatically. This will be addressed later but here we need to realise that these whole states are quantum states of (a whole) energy (oscillation) and the frequency increases with degree of how much is spanned, that is, the greater the coherent state, the higher the frequency. [* A single 'large' wave (picture sine wave) carries (spans) smaller waves of information of how much is spanned; this sine wave is a single whole quantum pulse, repeating every wavelength of oscillation.]

Thus the greater the whole, the higher the frequency; it has to be for the wave to carry the lower frequency. Now, as stated above, the wavelength acts as a unit—a single quantum pulse. This is utilised by the mind as a computer bit (basic unit for storage of one bit of information), but that one bit can contain many smaller bits within it. This will become clearer later. This means the computer bit-size of the greater whole carrier wave, is smaller (shorter wavelength, higher frequency) than the bit for the lesser part (lower frequency, longer wave-length)*. A smaller bit means greater differentiation (smaller space), and therefore greater independence capability. *[The larger carrier wave, apparently contrarily corresponds to a smaller (computer) bit. However, its size is spatial but its interval in time is smaller; that is, these are waves in time, and for the practical application would be scalar-standing waves. This needs to be pictured higher dimensionally, in a 4D direction; see Appendix E.]

There is a special relationship between integration and differentiation, which science hasn't recognised. This only occurs when the parts are in harmonic relationship, causing the phenomenon of quantum regeneration, mentioned above.

Thus the greater the wholeness of a harmonic system, the more detail it can express in the part. That is, the inherent wavelength is shorter (or frequency higher). The wavelength acts as a single whole; the smaller it is, the smaller the 'bit' size. When the parts are in a special relation, then integration is holographically related to differentiation. True unity is inherently holographic.

Science has not recognised that not only are basic particles whole, but so are larger entities, such as a planet, star system, galaxy, etc. (leading quantum physicist David Bohm suggested this some 40 years ago). These larger energy states underlie the particle level. The larger are these quantum states, the higher the frequency of their oscillations. A galaxy is a much higher frequency than a star system, and a planet is higher than an atom.

The main point we wish to make at this early stage is that the whole comes first and inherently has a higher frequency and smaller bit size, enabling more detail of physical movement to be controlled (density of information can be greater).

These greater quantum states of coherent energy and higher frequencies are not easily observed by science. The experimental method, utilising physical senses and scientific instruments, has a relatively low degree of order in our 3D and when interacting with higher orders (wholeness, integration), collapses the wave function, that is, quantum reduces the greater coherent states to the lower one—that is, to the parts (which are not in special relationship, or coordinated). This occurs without the observer realising (this is covered later, though mainly in volume 1).[1]

This is a difficult and technical area; it is not essential to understand it here. However, some understanding would give more conviction of the truth of the information, pursued here.

This argument, regarding the whole and the part, and which comes first, can be resolved by rationalisation and logic—though we would have to exceed the bounds of simple 3D logic. In fact it must ultimately be resolved by the mind; there can be no true basis or stable reference from the part, since the part has then to have its own context—some kind of background, and we start the infinite regression test-of-truth failure (Section 4). The beginning can never be found in the bottom-up principle (even by logical argument, which we are applying here). Everything tells one, including the tests of truth of physics, that only the highest order, the Absolute, can qualify to be the beginning. It is the final stillness and true zero point, relative to nothing, and passes the infinite regression test and, in fact, all the tests of truth in physics.

Let us summarise this section, in particular, regarding the relation between the whole and the part, utilising the following.

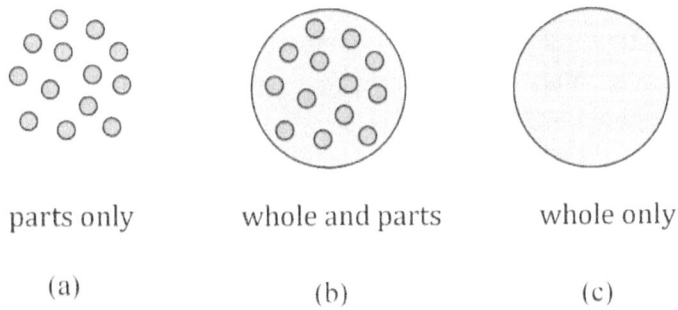

FIGURE 2: Three fundamental categories of all forms and manifestations.

We could say that there are three classes of form, or energy/-mass formations. In Figure 2(a) we have the parts only. These could be particles, atoms, rays of light, or even people. In (b) we have not only the parts but also the underlying whole quantum state of energy (like a collective). In (c) we have only the whole.

Note that the whole always has a higher oscillation rate than its parts (b), in fact, in natural products, such as atoms, laser light, planets, etc. The frequencies of the parts are sub-harmonics of the whole. In (a) we only have parts and so this class provides for artificial products, such as a motor vehicle which is built of parts stuck together by welding and nuts and bolts. There is no true whole in artificial products—the whole does not have true unity, only simulated unity, or composite unity. In Figure 2(b), this form provides for natural entities. A planet, or star, universe, etc., will have a whole quantum state underlying the atoms/molecules, which are held together in 3D by electromagnetic forces. The collective oscillation is slightly higher dimensionally. These greater coherent states (unity) are not usually detected by scientific instruments. Also collapse of the wave function occurs of the whole, leaving only the parts. Science will not detect the higher frequency wave (coherence of the individual rays of white light) of the laser.

In (c) we have the higher-dimensional wholeness (corresponding to that in Figure 2(b)); it is not available for 3D viewing. However, our minds are creating these states regularly. Good examples are in music and art.[2]

Of course, the wholeness is relative. Atoms are whole, containing sub-atomic particles. And particles, quanta, must have a size that is whole (a thickness in 4D); but this can be considered a uniform energy at our 3D level. In fact, as we shall see later, it is just a pattern in the aether, such as, for example, a tiny vortex, or whirl, for an electron.

Notes
1. Book: *The Emerging New Science,* volume 1.
2. Ibid.

3.

THE LIMITATIONS OF SCIENTIFIC METHODOLOGY AND OBJECTIVITY

Knowledge, and therefore progress and ultimately evolution, is retarded by the tacit doctrine that the only acceptable approach to truth is through the experimental method.

One of the principal weaknesses of the scientific method is its failure to model or even recognise unity; the wholeness aspect of reality. Science would do well to incorporate in its curriculum a study of the work of philosopher, Leibniz. Science avoids the issue of the significance of unity by considering it as merely consisting of parts held together by forces, forming a composite, or simulated whole. How did science fall into this trap of setting self-limitations? A primary cause of this is the following.

General programming on this planet has educated man to fall into the belief that the stability and accuracy required by science must necessitate the elimination of *all* subjectivity. Consequently within science the subjective and objective aspects of existence were separated. However, one cannot separate the two; they are two sides of the same coin. Nevertheless this can always be done artificially (see section on artificial realities). As a result, science endeavours to be totally objective, and correspondingly religion became imbalanced towards subjectivity, relying on faith and belief (also intuition, which is a valid subjective state). Thus science and religion separated out from what might be called a spiritual science, and each is the result of a partial quantum reduction from this higher truth (or context).

We can begin by showing that scientific methodology or the experimental method, which means proving theories by rigorously conducted experiments, has severe limitations that are not being taken into account. In doing this, however, we won't have actually shown that it is not the best method, no matter how limited science is. This is another problem. Nevertheless, to know how science can go astray is a great step forward.

There are two major impediments that are quite different from one another: one is difficult to understand but easy to believe, the other is easy to understand but difficult to believe. The first of these major problems is how far the experimental method can be successful. Quantum physics exposed this many years ago. The confusion was that radiation, such as light, was either made up of particles or waves, depending on how it was observed. In other words, what was perceived, detected or evaluated was in the context of *what* was making the observation. The missing key here in science is that all energy and knowledge are contextual. This is fundamental to the proper evolution of knowledge and of life itself. Everything that is observed is in the context of *whom* or *what* is doing the observing and from *what* viewpoint, station or perspective.

In the above quantum physics example it is the scientific set-up that is the 'what'. The 'who' also plays a secondary role in the form of physical senses (of the experimenter). The result of the scientific method is thus based on the scientific instruments and physical senses. So what was the conclusion of the wave/particle dilemma, mentioned above, regarding the contextual observation? The wave/particle dilemma arose from a long history of research and debate, starting with the study of light and so-called black-body radiation and the ultraviolet light catastrophe of which the only solution appeared to come from the notion that electromagnetic waves also behaved as particles. From this, it was then shown that in the two-slit experiment, when single particles, even electrons, passed through the apertures one at a time they still somehow interfered with one another, precisely as do waves. When not observed the electron could be anywhere, or everywhere at once. From this arose the question of nonlocality; the strange instantaneous link between two particles when separated over any distance. Einstein considered this spooky and decided that the condition of the particles was already known before hand, satisfying the nature of objectivity. Physicist, Bell, proposed an equation which posed this question clearly as to whether reality was objective or not. Remarkably, experiments were carried out by Clauser and then more accurately, Aspect, which proved the condition that the universe was not truly objective.

Thus the conclusion was that the observer (scientist and instruments) must be part of the experimental system and not truly objective to that system. But objectivity is the very basis of how scientific knowledge is acquired. The entire edifice of science rests on the principle of objectivity. Without objectivity the value of scientific measurement is in question; this is the way it has been set up.

This problem is scarcely recognised within science, and certainly not understood—it is thus not taught, other than to give the subject a passing comment. What does it mean for the scientific observer to be part of the experiment? Charles Darwin might have understood this better than most scientists today since he is known to have stated in a letter (and said in reservation of his theory of evolution, or anyone else's) that, 'How can man judge (or understand) nature if man is part of nature?' This was a very profound statement and not understood properly today. However, Darwin didn't recognise what the explanation was, or the solution. Much later, quantum physics came up with this same problem in scientific methodology, as mentioned above. Note that one of Einstein's philosophical wisdoms was to state that in the observation process, consciousness must be viewing from a higher perspective than that of the observed.

This is all to do with references. In life, also nature and the universe, references are informational levels (often fractal levels) that are knowingly or unknowingly taken as a zero—something from which measurements and judgements are made, etc. These are *contexts* (recall that all knowledge and energy are contextual). When the observer is inside the context (part of the system), such as in the experimental method, the boundary of the set-up is automatically and unconsciously taken as a zero (when in fact it will be found to be part of, or inside of, a larger system). This means that science is not measuring true physical constants but only relative values—nevertheless they are constant over a certain known range, such as a fractal level (say, just 3D and not the expanded dimensions beyond this).

A consequence of this is that it also means that most of our physical laws are not immutable (constant), and that the limitations that they impose on us and on the universe are relative. In particular, this means they can all be bypassed. Einstein's relativity is an ingenious representation of the illusions of the third dimension, but this is with respect to its own context. It is correct as per (in the context of) the observations and measurements made. Anything will self-prove relative to its own context. It may not be true relative to a wider context. One of the keys to the foundations of knowledge is the relative zero, which is not recognised. If that sounds too mathematically abstract we can use the word 'context' (instead of relative zero; see Section 5).

When a context is unconscious or unknown, its role as a reference point is assumed (unconsciously and incorrectly) to be a zero. A simple example would be the prejudiced mind—it does not know that its apparent rational part of its thinking is unconsciously being biased, causing failure in clear thinking. See Figure 3, page 46.

We have examined the word 'what' (for example, scientific instruments) in the expression 'whom or what is making the observation'. The 'who' in this case of scientific methodology is the physical senses. The perceiving and detection system in the experimental system is achieved by scientific instruments *and* human senses. Both these are within the same frequency band as the environment that is being measured, and thus are part of, or within, the experimental set-up.

In order to evaluate something the observer must be outside or 'above' that which is being observed (e.g., can't see the woods for the trees; therefore step outside the woods). However, by 'above', in this context, we mean occupying a higher state of resolution. This different frequency level can be explained by recognising that one wouldn't use a camera with a film that has a grain size large relative to the items being photographed. Of course, in the experimental set-up, this is governed by the physical senses and equipment limitations. Other ways of acquiring truth must be introduced into scientific methods but this must inevitably require more assistance from right-brain intuitive faculties, which are not being given full reign.

Nevertheless, we may still ask why the observer is part of the system being observed. It is not just a matter of frequency, such as radiation, but also the coherent structures that the higher frequency creates. The universe functions on a system of orders (degrees of coherence); it is a fractal, holographic structure. The level of order reveals the true basic meaning of intelligence—the degree of order (coherence, unity, integration, harmony, coordination, interconnectivity, etc.). The experimental set-up of scientific instruments, and using physical senses, is of a relatively low order. It can only detect (select or resonate with) its own level of order. This applies to all observing systems, instruments, consciousness in different forms, and even energy (frequency/wave) patterns. An observer of the same order as the observed could detect the parts that compose the observed order, but will not detect the whole, and will collapse the wave function down to the part level (where we have parts stuck together by forces).

Regarding the two analogies mentioned, that is, 1) can't see the woods for the trees, and 2) using a camera film with a large grain size relative to the items or details being photographed. The latter (2) handles the frequency aspect; a low frequency system (large 'grain') can't detect a high frequency system. However, this is just a problem of radiation. In (1) we may have a coherent structure being observed, consisting of the same frequency values. Why can't its presence be recognised? The same frequency could be detected but if the frequencies form a coherent structure, this greater wholeness quantum regenerates a collective for the whole, which is a higher frequency—recall Section 2. This is a higher order than the parts that make it up—the coherent whole is greater than the sum of the parts. The lower order will collapse the wave structure of the higher order, as per the collapse of the wave function, and the higher knowledge is lost. We shall give some examples later as we deal with this subject of the 'collapse of the wave function'; probably the most important phenomenon in the history of science.

The 'observing' context 'collapses' the wave function to give data commensurate with the reality of the interacting 'viewpoint'. Thus scientists are unwittingly quantum reducing higher-order (higher truth) aspects of the universe to correspond to their own contextual order.[1]

Let us now take the second major stumbling block; much easier to understand but difficult to believe. We can begin with a simple analogy. Imagine a group of investigators encounter an old motor car, but that they have no knowledge of such a machine—in fact this machine is being used as an analogy for a region of the universe or environment to be researched and understood. They proceed with their scientific evaluation—we can have the engine running to help the analogy. They might note and record the following (upon which new scientific principles might be based): six cylinders, say, four cylinders active and the remaining two cylinders inactive or 'blank'. Let's say they consider that this is supposed to be how the engine functions (only four cylinders firing), and they formulate physical and mathematical principles or laws based on this incorrect conclusion. Similar, other malfunctions of the motor car are taken to be correct and natural.

We are actually implying that regions in our local universe (Milky Way galaxy) are similarly malfunctioning. Clearly if laws and principles of science are based on this, wrong conclusions will be drawn and further, since there is a symbiosis between science and consciousness, there will be corresponding distortions within consciousness, affecting genetics and subsequently evolution.

Our galaxies, in general, which are primarily dual-vortex systems, have a balanced white hole, black hole system. This has a frequency of oscillation from one pole to the other pole. This is normal. However, our scientists have discovered that our Milky Way galaxy (and some others) has a predominant black hole at its centre, and consider this as typical of all normally functioning galaxies—they may find out it isn't typical.

However, science recognises some flaws in the universe, in particular, that our solar system is unexpectedly chaotic with wrong orbits and spins of planets, etc. Also Earth is tilted approximately 23.5 degrees, which is not exactly a harmonic mathematical proportion.

Probably the most important, well-established misunderstood mechanism of life is the DNA. Over 90% of the DNA in all species has an absence of base pairs (humans over 95%) and subsequently is considered 'junk' DNA. Unfortunately when the first geneticist discovers what the junk DNA is they will likely be ridiculed, attacked, and it will be a bigger cover-up than the UFO scene, since it will alter our whole view of the history of evolution and consequently the prospects of our future evolution.

Further, the particle and antiparticle annihilation is not understood, or perhaps we should say misunderstood, due to the above-mentioned impairments in some physical processes. Where nature is functioning as per its original harmonic blueprints the particle and antiparticle combine in the evolutionary process to form a unity—a higher frequency particle, which then merges into the next inner-fractal level as it divides again into two new particles related as per the new fractal dimensional level. This particle interaction is fundamental to evolution.

Scientists, however, do encounter in the laboratory a different outcome, in particular, that the particles annihilate one another as matter and antimatter in a burst of energy. This automatically assumes a wrong zero or context from which 3D processes are measured. Whereas when the two particles combine to form a greater unity, this merges into the next fractal level, which is a new context or zero reference (until the latter is shown to be part of a still higher context). Thus there are polarities (like positive and negative, and particle and antiparticle) within polarities, etc., in a holographic system. Subsequently science again becomes misdirected when endeavouring to formulate absolute laws, that is, forming ones more global and universal. Current science is merely discovering laws pertaining to local and relative contexts only.

Science only recognises the first fractal layer, like detecting only the twigs on a tree and not being aware of its fractal connectivity through increasing integrations of branches to the single source, the trunk. The twig is carried by and is within the contexts of the higher-fractal branches, as is our 3D 'carried' by 4D, 5D, etc., which are dimensional spectra or wavebands.

Thus these are the two main impediments to scientific discoveries or knowledge and invention, which will subsequently limit human abilities and steer evolution off course.

We see from Figure 1, Section 1, that science completes itself with one 'material' level of creation (3D), the lowest spectrum in Figure 1(a). This means life, people, the human observational viewpoint or the observing scientific experimenter is also functioning on this same frequency or wave band (even referred to in quantum mechanics as a participator in the process of creation). A proper evaluation of a system cannot be made while inside that system (on the same level, which means within that context). A machine can't evaluate another like machine. This is what puzzled Darwin, mentioned above, when he stated in that letter (which was read out on a radio programme), 'How can man judge (understand) nature if man is part of nature?' His question is based on the assumption that existence is made up as shown in Figure 1(a)—man being part of the same substance as that which he is observing.

A very good analogy for this is the well-known saying, 'Can't see the woods for the trees,' to which we have already referred. One has the viewpoint inside the woods, as does any of the trees, and can't see the boundary of the woods or the relationship of the woods to other woodlands. One must be outside the context to evaluate it (for example, above the woods). It wasn't realized by Darwin or scientists, in general, that the mind extends into higher frequencies and can thus always (potentially) attain an observational state higher than nature or the universe, as we might see from Figure 1(b) or (c). Consciousness and mind extend with their higher aspects up the frequency spectrum and thus can potentially evaluate any context by going into a higher viewpoint. This viewpoint, or point of observation, is governed by the resolution level, which is wavelength. The mind is always capable of going higher in frequencies (higher resolution, shorter wavelengths) than the universe region being observed.

The New Science handles this adequately but does quantum physics help here? Indeed, it is in fact one of the main sources. Quantum physics exposed this experimentally some 80 years ago and drew the same or similar conclusion as did Darwin. In this case, it was that the observer (experimenter) was part of the experimental system (compare 'part of nature'). Note physicist John Wheeler's statement regarding the experimental method *or* any observation, 'The man who stands safely behind the thick glass wall and watches what goes on without taking part . . . can't be done'.[2]

This is saying the observer is not truly objective in our scientific methodology—thus the basis of the acquisition of reliable knowledge in science is flawed. Why is this? It is because science only uses physical senses and scientific instruments, which are structured within the same spectrum of frequencies as the apparatus (experimental environment, nature, and the universe).

As stated, quantum physics revealed that, 'What is observed is in the context of whom or what is doing the observing.' Think of the observer as either, or both, a scientific instrument and the human as an organic robot (a machine). But the human has actually a higher-order mind structure (with the inherent absolute property). Science dictates that only the scientific instrument and the robot aspects (for example, physical senses) are used in the observation, all of which are in the same class (order) of matter, energy, space, and time as the observer (environment). This is what is meant by being part of the experimental set-up. The mind must be developed and ultimately involved in order to 'step outside' of the system (connect to the higher aspects/fractals). The more complete explanation, however, is revealed when we use the quantum physicists' expression, the 'I/not-I', instead of the observer/observed. This will be better understood in Sections 10 to 12.

In summary, quantum physics was running into the relationship between mind and universe; a subjective element had been revealed which was universal. This was an almost unconfrontable shock to orthodox science since it meant mind and universe are intrinsically interconnected (that there is no objectivity); meaning we can't have one without the other once the universe is manifested, but of course the subjectivity existed first on its own. This means that in current thinking, because the observer is made up of the body and its senses, and the scientific instruments, the observer is of the same substance or degree of order as the observed; precisely what Darwin sensed was a problem with (physical) empiricism. This was corroborated by quantum physics, with a further understanding provided by the 'I/not-I' relation. However, Figure 10, in Section 10, indicates there is a whole scale of observers/observed. This means that the mind must be developed and used. A higher-order (mental) empiricism must be used. The next level would be the 'soul' input in the body, providing a higher-order observation, and the body and brain become the observed/environment, which is now the unconscious mind.

It may be of interest here to mention that invisibility, or the undetectability of energies or objects, occurs when the energy is of a higher frequency, and also when it is oriented differently, not 3D-wise, but in a 4D direction. The military 'Stealth' invisibility goes beyond the screening of radar (not publicised). Nevertheless the invisibility is created by electromagnetic force-fields, which bend light (around the object). The object is still there though and detectable by more direct means, such as physical contact. This was the basis of the Philadelphia experiment in 1943, but in fact surpassed the electromagnetic invisibility (due to intervention from other factions).[3]

In general, however, where the scientific method is applied to observations and experiment, which are relative to a known context, such as measuring the acceleration due to gravity, there is no problem; results are relative to a known context and are correct. But where science is pushing the boundaries into new territories, the relative aspect becomes more important, and the scientific method will only give the lowest relative results; for example, physical constants are only relative and will change with expanded contexts of knowledge. They need to be updated periodically, which would occur automatically with a proper expansion of knowledge. This scientific limitation will inherently restrict knowledge and subsequently adversely affect our evolution. Simply proving something without the qualification of context is not absolute proof: anything will self-prove relative to its own context.

The answer is not simple. Scientific methodology should continue to be used as much as possible but keeping in mind the limited context. The educational system should emphasise more right-brain abilities, intuition, imagination, inspiration—and there should be acceptance of a mode of acquiring information from many sources. Eventually with a powerful development of the intuition, research will move in the most advantageous direction and experimentally untested theories could be evaluated for truth content by the intuition. This is governed by the degree of (universal) subjectivity compared with objectivity.[4]

In a nutshell, an advanced civilisation has no need for the experimental method; the members can use their minds to acquire scientific data. Everything can ultimately be known through the mind and consciousness development, achieved by consciousness 'projecting' into that which is to be understood, since everything is made up of units of consciousness.

Thus the experimental method only gives first-order relative results. For example, consider the non-classical aspects of Einstein's special relativity. As per any and all rigorous scientific measurements at the 3D level, the velocity of light comes out as always the same, forcing science to contrive unexpected and unnatural distortions in space and time relationships to satisfy the observed constancy of the velocity of light. The reference level is 3D. If we were aware of the extended spectrum of light we might find that the velocity of light is not constant.

More attention should be given to 'tests of truth' (originating within physics), which actually have a higher rating than the experimental method itself. See next section.

Notes
1. See book *The Emerging New Science* for a more in-depth account of this.
2. Book: *Space, Time and Beyond* by Bob Toben.
3. Book: *Engaging the Extraterrestrials* by N. Huntley.
4. www.nhbeyondduality.org.uk. Articles: *The subjective/- Objective Illusion of Quantum Physics* and *The Limitations of Scientific Objectivity*, and the book: *The Emerging New Science*.

4.

THE IMPORTANCE OF TESTS OF TRUTH IN PHYSICS

A qualitative evaluation that indicates the level of truth of a theory and which has a higher rating than the experimental method that only gives relative results.

'Tests of Truth' are just that; an application of logical or even common-sense procedures to determine the degree of validity of a solution or theory. They enable an evaluation to be made of the degree to which the data succeeds as an explanation. For example, symmetry is a popular and commonly recognised indicator of truth. But the most important feature, as we shall see from the examples presented, is that the direction of truth runs parallel to a scale of decreasing relative values, but increasing absolute values.

The example we gave above, *symmetry*, is an easy one to understand. We hear of beauty in mathematics or even physics laws, etc., in particular, there may be a certain symmetry in the equations. There are other types of symmetry, which may come under *preferential formats* (see below). For example, let's say that one scientific investigator arrives at the nature of a basic particle and evaluates it as cylindrical in shape; but another researcher finds that it is spherical. We can see clearly that the sphere has the symmetry; the cylinder has a preferential axis. The view of the cylinder changes as we turn it around; but the sphere remains the same from all viewpoints.

The sphere is more absolute. Ultimately symmetry brings about the elimination of stray and dubious factors that do not integrate or unify to achieve an order higher in the fractal scale. However, one must beware that this elimination does not possess intrinsically another higher-order symmetry, which is then suppressed. This may be vital to further developments of a theory. Let us elaborate further.

An additional point with symmetry is that when it applies to theory, the scientist or mathematical physicist, having established the satisfactory practical results from a symmetrical formulation, generally will not recognise that this will usually be symmetry relative to a given context. What happens now is that another investigator may eventually observe an anomaly in the theory that it doesn't explain. The human ego will resist this additional information since it upsets the symmetry or introduces more confusion. It may be wise to investigate this anomaly as it may lead to further anomalies, which then regroups the symmetry and expands the scope of the theory to a new higher context—a greater truth and greater symmetry—that is, a higher order.

Let us present scientifically-familiar tests of truth:

Avoid *infinite regression*.

Avoid *dualism*.

Avoid *preferential formats*.

Apply *generalisations*.

Apply *covariance*.

Apply *symmetry*.

The question of *infinite regression* arises when we theorise about the beginnings of creation. If one can keep asking, 'What came before that?' Then the theory is inadequate or incomplete. All scientific theories of creation fail this test (since they are 'bottom-up' theories). Ironically the religious God concept would be a 'top-down' theory and passes this 'test of truth' but is unsatisfactory to science, since it is based on the subjective and the experiential.

We have seen in the fractal system, increase in order is increase in integration, and an intrinsic factor of learning is interconnectedness. This is clearly higher in the evolutionary scale. Thus removal of dualism achieves a state nearer the Absolute. Infinite regression forever searches for the absolute reference and never reaches this final reference. An ordered fractal system is a sequence of relatives/references/contexts of increasing degrees of order culminating in the final reference, the greatest whole, which we can call the true zero from which all other relatives are measured. See next section.

Dualism, arising in a theory, refers to two states that appear incompatible, and therefore cannot be considered to form a satisfactory coherent theory. For example, spiritual and material, where one is simply considered nonphysical and the other physical, without further understanding. Imagine a fine mist of water vapour invisible to our senses. A primitive view might be that there is nothing there. In contrast let us take a block of ice. We might say this is dualism, not realising the connection between the water vapour and the ice, that they are basically the same; simply condense the vapour and freeze the water. Thus the apparent dualism is handled. This means a theory that has water vapour and ice in it has less truth, is a lower order, than a further development, a higher context, which reveals the relationship between the vapour and the ice.

Dualism also applies even to very reasonable and scientifically acceptable theories, meaning eliminating the dualism brings a higher relative truth. For example, in Einstein's general relativity, gravity is explained by the effect of a mass, present in space, warping that space, and this can be interpreted as gravity. This is a better theory than offered by Newton's corresponding law. Nevertheless, it doesn't have the higher merit that quantum theory indicates. In general relativity we talk about a mass, for example, the Earth being placed in space and causing curvature of space (the warping effect), but note that this model has two very different entities, space and mass. This is a level of dualism: two totally separate characteristics without a common denominator.

Now take a look at what quantum physics offers. This can be called the ocean model, proposed by quantum physicists over 80 years ago. Quantum physics says, imagine a tiny whorl (whirlpool) on the surface of the ocean, that is, water spinning inwards and away (down) from the surface—this has the characteristics of an electron. Note the importance of this. The particle, the electron, cannot be plucked out as a distinct and separate particle. It exists entirely on the basis of moving water plus shape of water. A typical quantum physicist's description of this phenomenon is a quote from Shroedinger (probably the leading name in quantum physics): '. . . all bodies and objects [this means all particles and in fact all manifestation] are like shapes and variations in the structure of space.'

Nevertheless an earlier form of this model arose as a result of electromagnetic waves being detected in space but not the medium they wave in. The physicist Hertz demonstrated sending radio waves from one room to another. However, at that time space appeared totally a vacuum. Therefore a medium was postulated, called the aether, to provide a medium for the electromagnetic waves to be transmitted. Now we see that the ocean model provides a simple visible picture of this. And further, we see that the concept of a single medium, water, explaining both medium and mass, is a higher truth than the case with general relativity. The dualism has been transcended.

Generalisation is a particularly powerful and obvious test of truth. The more a theory, or a principle or law, explains *more* phenomena, that is, it generalises further, the greater the truth it embodies. As alluded to earlier, it indicates this direction, which is towards the ultimate answer within the Absolute.

Covariance can be a tricky one. The Absolute is automatically covariant; it 'looks' the same from all observing perspectives. Objects move in space, that is, in a background. This background needs to be the same relative to all observers and not have preferential biases, such as an object drifting in this backdrop at a different relative velocity (to the background) from another object. This was the problem when scientists proposed the aether to provide a medium for the transmission of electromagnetic waves. From the viewpoint of the Absolute all bodies are still (the Absolute, in effect, accommodates all velocities); the reason is, that it is the *same* absolute that is in contact with bodies at any time; remember the Absolute transcends space and time. Its relationship with all bodies remains the same (a difficult concept).

When the aether was postulated to provide a medium for the newly discovered electromagnetic waves, not only could it not be detected by accurate experiments (Michelson-Morley) but a theory couldn't be devised which would enable the aether to pass its associated test of truth. That is, the aether must be covariant relative to all bodies. It must have the same values relative to the viewpoints of any moving bodies—like the Absolute.[1] See Section 21.

Probably the most overriding test of truth is inherent within the relative zero; meaning how far is the level of truth in the range, relative truths to the Absolute. Inevitably we must look into self-referencing systems—see next section.

Notes
1. Article: *Superspace and a Covariant Aether.*
www.nhbeyondduality.org.uk.

5.

SELF-REFERENCING SYSTEMS AND THE RELATIVE ZERO

Zero is not only a number but is the most important number; and moreover in its applications it is relative.

Although we haven't yet introduced consciousness (see Part 2), it might be helpful at this point to reference it, since we have seen that observations using the experimental method are contextual, not just with respect to the scientific instruments but also the consciousness of the human observer. Thus at some point we must include this counterpart to the universal fractal system.

Consciousness can be observed to operate in quantum states, meaning it manifests whole, holistic, gestalt perceptive frames, whether there is a sharp point-like focus or there is a uniform spread. Moreover, these states of consciousness rapidly switch from one focus to another, giving the impression of continuity, time and the presence of gradients. Nevertheless, consciousness can become fixated, or be specifically hypnotised. It then forms a closed, self-referencing system. This is what we are interested in: self-referencing systems. The parallel concept for this in technology, that is, machines, is 'closed systems'. We shall come back to this. The designation 'zero' tends not to be regarded as a number, even in mathematics. In fact, zero is not only a number but is the most important number and, moreover, is relative; meaning, in general, it has relative values.

Where does the idea of zero come into all this? In effect, a self-referencing system has given itself a zero context at its boundaries; that is, at its interface with any greater system within which it resides. In fact as far as the self-referencing system is concerned there is *nothing* outside itself—hence the use of the term *zero*. When we design a meter to take measurements we put a zero on the scale which acts as the reference for the quantitative evaluation by means of numbers on the scale or dial—say, for example, 4 or 5 units of measurement. We always set the pointer to zero before measuring and we then assume this is an absolute zero. Clearly it is zero for our immediate reality but it will only be relative to the environmental set-up. For example, are the physical constants of science really constant? Scientists will find that even Einstein's relativity is part of a larger context, but it is set up as is all physics and mathematics so that the self-referencing system appears to be the only one—the context is closed off, and anything will self-prove relative to its own context.

For simplicity we can take the Newtonian relativity of motion to use as an analogy in demonstrating relative zeros. Let us consider a train compartment. In this case we have a smooth running coach and the blinds are down. Imagine there is no evidence for not only motion of the carriage but no evidence that there is any existence beyond the interior of the compartment.

If we measure velocities of, say, simple hand motions from side to side we will think these results are absolute (remember, this is only an analogy and 'velocity' is a suitable metaphor for many other types of measurement). If we wish to take the velocity measurement a little more literally, it means we have measured the velocity relative to the boundaries of our system, that is, the compartment that we take as zero. (Note that in our real universe we normally measure velocities relative to our planet, which is taken as stationary, or measure our planet's motion relative to 'fixed' stars.) In the train compartment we do not realise that the boundary of our system—the walls, floor, roof—are an interface to a larger system which we may peek at for a moment through a crack in the blinds. Clearly if we measure the velocity of the same hand motion from the larger perspective (outside the train) the result will be different. Thus the first result was relative and not absolute. Similarly, physical constants of science will be found to be relative values to roughly our 3D environment.

The reality within the train compartment is a lower 'fractal' dimension and the greater 'outside' reality is a higher 'fractal'. Are there more? Yes, it would not be logical to consider only two, but we are getting ahead. More examples must be given to enable the reader to grasp this subject. As a contrast, so that the reader develops a broad view of this concept and assimilates the extensive, in fact, all pervasive scope, of this concept, let us give a qualitative example.

We can all perceive when another person is prejudiced or biased, since we are outside the 'system'. But the prejudiced person is dramatising a self-referencing system and can't see this. They generally can't step outside of themselves to reset the program; that means connect the data up to a new context, which is the greater perspective in which the irrational relationships will be observed as obvious. The person is in a kind of hypnotic trance.

Consciousness, however, always has the potential to step outside itself (the lower level), since it is fractalised, as is everything natural—see sections on fractals. The self-referencing system in the above example of the prejudiced person will zero out the bias (their conscious viewpoint is resting on the bias but *not* including it; see Figure 2. This means the person will be unconscious that their evaluation (behaviour) is based on unconscious irrational data. In the case of the train compartment example, the contribution in the measurement made by or provided by the larger context, such as the outside (the train) view, will be zeroed out if the occupants of the compartment are only aware of the interior. The self-referencing system will judge all unknown contexts as zero. Let us give more analogies. The see-saw example is a good one. Imagine a plank balanced on a fulcrum with a person seated on either end. The characteristic motions of the plank are governed by the fulcrum—this is the zero point. If we consider measuring the amount of rocking of the plank, that is, the amplitude, this measurement references the context of the still point of the fulcrum. We can imagine this is a complete and closed system, such as this third dimension or even an environmental set-up in which rigorous scientific methodology is being applied. The results will appear absolute. But if we now realise that this first fractal system, the plank, fulcrum and persons, is on the end of a larger plank and fulcrum, with persons on the other end, then our new zero reference is the second fulcrum. The first fulcrum is now moving up and down since it rests on the end of the larger moving plank.

We see that the result of the measurement made in the first system will now be different when we reference the context of the second system and the new zero (fulcrum). Thus the first zero was only relative; similarly for the second fulcrum, and so on to a complete system. The fractals do not continue indefinitely just as a twig on a tree is connected to a larger branch, which in turn is connected to a still larger branch and so on, but completes with the full integration at the tree trunk (and in life and the universe this becomes an absolute—see Section 6).

On the subject of trees, a further example could be a bough of a tree moving in the wind. The twig on the end traverses the longest arc. Let's call the twig the first fractal. Then the twig is connected to a branch—the second fractal—which in turn is connected to a larger branch, the third fractal. Each incrementing fractal moves less than the previous one. Thus we could say that the twig is not 'aware' beyond itself, that is, aware of the second fractal and higher. It will regard its motion of the twig relative to its connection to the second fractal, which it takes as zero (motion). Thus the twig in this analogy only experiences its own minimal movement but which in fact is being carried by the accumulation of the other motions of the branches.

The human arm is a similar system. The joints give us the fractal levels. If we imagine arm motions where all joints—shoulder, elbow, wrist and fingers—are active, we know from results achieved in experimental psychology that the fingers have their own learning patterns; the wrist has its own learning pattern, *plus* the fingers; the elbow has elbow, wrist and finger learning patterns, and the shoulder joint has them all. Thus the fingers, say, moving on a keyboard, have their own closed self-referencing fractal level. They can only move up and down and are not 'aware' of being dragged to one side by, say, the shoulder action. Learning *can't* pass up the hierarchy from fingers to shoulder; only down.

The twig is also a good analogy for human consciousness, which is connected to the second fractal level, the 'soul', and so on. The human consciousness is only aware of its third dimensional activities, and not its greater fractal selves. Owing to excessive negativity on this planet the connection between human fractal and soul fractal is weak and thus the human consciousness is extremely vulnerable to the creation of a closed, self-referencing system. Thus, in general, conclusions tend to be based erroneously on 'absolute' facts, meaning they are not actually absolute.

Let us give a simple example of a relative zero, for example, in electronics. Draw a graph of a sinusoidal wave motion; see Figure 3(a). Note the voltage intensities oscillating from plus 5 volts to

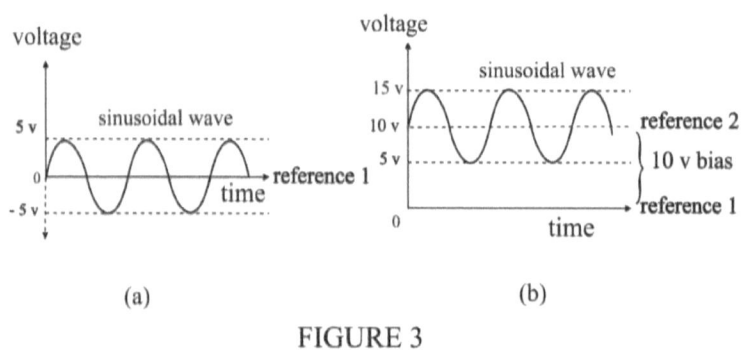

FIGURE 3

minus 5 volts. The reference for this wave is the neutral line, the zero. Now in Figure 3(b) we have given the wave a bias, say, 10 volts. Now the wave oscillates from 15 volts to 5 volts. Its new reference (that it rests on) is 10 volts.

As an analogy, if in Figure 3(b) we are still operating in the context of Figure 3(a) the 10 volt level can appear as zero. Within the context of Figure 3(a) it will work but we see that if one expands the context to include the bias it gives a different result. For example, atoms are in the context of, say, a cell, of which the cells are in the context of an organ, this then references the body, and the body references the planet, star, galaxy, etc. This simple means for instance the atoms have their own individual characteristics but in the context of the cell are influenced by the whole cell, which in turn is influenced by the whole organ, etc. There are wave patterns resting on wave patterns, repeatedly. The interconnectedness of all things requires that a complete reference system of contexts is present.

We may now see how our meter may only be referencing a relative zero. Our objective scientific methodology creates a closed, self-referencing system. The observer/observed relationship is an interdependent one—it is context-dependent. Only relative results can be achieved. Our so-called constants are merely values which reference the interface (boundary) between one system and the next, which because there is no awareness of the larger (next) system or dimensions, the reference is automatically zeroed out.

The formulation of scientific data reaches its limit in the very objective set-up it creates. The conditions of objectivity set up an experimental reality involving devices and experimental arrangement, designed on the basis of objective knowledge; all with emphasis on 'seeing is believing'. Such a set-up predetermines what nature will reveal to us; meaning it is governed by our state of objectivity and unconsciousness.

This system will self-reference, forming a closed environment. In the previous analogies, compare the first see-saw, referencing the fulcrum as a zero or the twig only 'aware' of its own movement. The more objective we make the elements of the experimental environment, the more consciousness denies its participation, and the more the unconscious factors, such as higher contexts, are greater than the conscious factors and close them off. Consciousness is then connected like the dog chasing its own tail. It creates a closed system. To reveal perpetual motion one must create knowingly, or accidentally, a 'leak' out of the system. The first see-saw with its limited, relative data is violated by finding a link out of the first system to the next (fulcrum). Objectivity establishes unconsciousness; see Figure14. In a fundamental sense the environment is separate from us to the degree that we are unconscious of it (but this environment is part of the extended unconscious mind). See later sections.

This level of scientific methodology will deceptively give us the constancy of the velocity of light, Planck's constant, the conservation of energy, etc. but will automatically fail to incorporate features outside the first system. Ironically, the more positive and thorough the experimental set-up, the more it will be limited to its own context and not even a glimmer of the next fractal level will be present. However, in using the train analogy, if we get a glimpse of the passing environment through the gap in the blinds, our internal observations are changed since we have, in effect, stepped outside the system. Under these new conditions, intuition and imaginative states of mind interacting with the first objective environment may cause a different selection of data to occur (a change in selection of probabilities), and the experimentalist may find he or she obtains a different result, which note, may to their great frustration not be reproducible. The 3D objective environment was unknowingly modified and now unfortunately the 'desirable' hidden influence (on the first fractal from the greater system) removed; that which is not part of the established formalism.

For example, the non-graduate scientific pioneer is more likely to design a set-up in which a normally closed Newtonian, non-perpetual motion device has a flaw in it in which certain unknown elements of the design are manifesting an open-system characteristic, with the subsequent anomaly of, say, the production of extra energy—and the potential for perpetual motion.

The basis of this is that the environment changes according to the spectrum of consciousness. The subjects of psychotronics and radionics are examples in which because of the above reasons, results may not be reproducible by all investigators—often the same person. In some circumstances, the conscious observation itself may cause the system not to work in that way. It is easy to see why strict objective conditions stifle ESP and paranormal energies. If the activity is allowed to operate without analysis it may work. Paranormal energies are outside the first (3D) system and our stringent, scientific, objective conditions determine the frequency range—the lower one. Thus science fails to detect the higher frequencies, but also where there are coherent structures, collapse of the wave function will occur (see Section 3 and later sections). Similarly, understanding of art (and music) can only occur when analysis is not taking place—and in fact when the 3D attention is out of focus. Such conscious and intellectual analysis will quantum-reduce the holistic (unified) aesthetics of the art to the particle level—phase correlation to phase randomisation, to use a quantum physics expression.

When we set objective standards we create a 'zero' from which to judge anything or measure anything. Collective agreement, programming and naturally-imposed limitations set the 'zero' and we don't know there is already a bias in our judgement. We could take a negative situation, say, a criminally-inclined society, and use this as an objective standard to evaluate the merits of behaviour (meaning accepting this society as normal). This becomes programmed and embedded. As stated previously, when programmed there is a self-referenced system that zeros out the bias or prejudice (from the viewpoint inside the context) and everything seems normal. We could program an individual to be triggered by a key phrase to go on a shooting spree—the question of ethics, morals and destruction will be zeroed out by the individual. Whenever we judge a standard we put a 'mark' on an imaginary gradient scale ranging from 'good' to 'bad', and use this as the context, the zero, by which to make the judgement. The point chosen (usually subconsciously) is a relative interface between 'good' and 'bad' and is a zero mark for judgement of how many degrees of 'bad', if it is on the 'bad' side of the mark, or how many degrees of 'good', if it is on the 'good' side of the mark (this causes the so-called 'double standard', which is actually often part of a multiple standard).

Returning to the problem of the closed systems of scientific methodology (which note is the (cost-) free-energy field), are we therefore saying that experimental conditions *shouldn't* be rigorous—that one should be sloppy with the design, etc? In our present condition, there is no satisfactory answer. We have strayed so far from a correct balance between subjectivity and objectivity and the symbiosis between knowledge and consciousness. As knowledge expands, it should not have veered unduly towards the objective but retained an ideal ratio with subjectivity. By this, we mean there should have been an increasing recognition (in evolution) of the role of the mind, and its power and participation in the environment. Now we have the unsatisfactory situation in which a pioneer with lack of strict adherence to scientific and experimental protocol is more likely to make the really valuable discoveries. Note that when they are made they are suppressed.

Science hasn't established that there are (macro) dimensions within dimensions forming a hierarchy of frequency spectra (basically scalar waves). It is a little like living in a box, which is the first fractal, such as our 3D, but which is within the next fractal, a larger box and so on. There must always be correct alignment between adjacent fractal levels, between contextual strata, otherwise chaos will result. Our sector of the universe appears to be designed on the basis of twelve main divisions or dimensions. The DNA will be found to have the potential for 12 strands, and we have many existing 'twelves', such as the notes on the musical scale and many more in religious history.[1] However, note that we should have been using base-12 for mathematics, and not decimal, that is, base-10. This itself will cause results, significant numbers, physical 'constants', of science and mathematics, to be out of alignment with our environmental dimensional structures. The '10' configuration will not fit into the '12' of nature for a natural resonance and synchronicity, and for a proper progression of discoveries towards a harmonic science. Similar distortions in the mind will cause chaos and insanity.

Let us finalise with another analogy, in particular, to show us a mechanistic interpretation of synchronous activities, such as the above-mentioned alignment between dimensional contextual strata, which also shows how the interface between different levels or loops should zero out for correct alignment. We have the example of the production of the television image (using the old cathode ray tube). The electron beam of the cathode ray tube is focussed into a sharp point. It strikes the screen at the top left corner, creating a fluorescent spot of coloured light, then moves across, horizontally. When it reaches the right side it jumps to the left just below the previous line and continues to do this 615 times (U.K. system) until the whole screen has been traversed. Then at the bottom right corner it returns to the top left corner and begins again. One may notice that we have 615 smaller loops—the horizontal lines—inside a larger loop, which brings the spot back to the beginning. It may be seen that the points of return must be precise. At the end of the lines there is a synchronous pulse applied which returns the spot for each line, and then there is a second synchronous pulse which returns it for the whole picture. If these pulses were out of phase the picture would go into chaos. The loops must be zeroed, that is, aligned at the precise time.

Since everything is made up of flowing energy and under continuous creation, it must be precisely structured in the above manner, whether it is coordination of the fractal joints of the human arms, or the dimensional organisation of the universe, following a base-12 code.

Notes
1. Information that the universe works on base-12 is available from workshop DVDs; transmissions from the Guardians by A. Deane.

6.

THE ABSOLUTE

*An 'unmanifest' condition of no motion,
particle, wave, or space and time;
everything else is relative to it and springs
from it. All of it is everywhere at once.*

The Absolute is also referred to as Source, the Origin or Primordial Cause, the Ultimate Reality, God, the Divine, Nirvana, Supreme Being, the Unutterable One, the Unmanifest and many more conceptual descriptions; also that it is an unconditional reality which transcends our limited, conditional existence. All these terms describe the source of, and force behind, all that exists.
Author I. K. Taimni describes it as:

> ... is the very core of our being as well as the cause and basis of the universe of which we are part. . . is beyond intellectual comprehension, still, from the intellectual point of view it is the most profound concept in the whole realm of philosophy. The fact that it is called 'Unknowable' does not mean that it is beyond the range of philosophical or religious thought and something on which thinking is impossible or undesirable. The very fact that it is the heart and the basis of the universe should make it the most intriguing object of enquiry within the realms of the intellect.[1]

Religious and philosophical doctrines embrace both personal and non-personal attributes of the Absolute in that it (the Absolute) may or may not be endowed with specific intelligence or personality. Nevertheless, it is the Source through which all *being* emanates.

Clearly the foundations for a science need to go beyond philosophical and religious interpretations of Creation and existence. The term 'absolute' has been referred to tentatively within physics in the attempt to understand the aether, more specifically, the covariant property requirement for an acceptable theory. However, in terms of science, no one knows what the absolute reference is; some think it coincides with the aether frame, but no one knows what the aether is. Einstein decided the absolute frame could be avoided, which is what happened to preserve relativity. As we shall see later the aether can have the necessary absolute characteristics but still be quantifiable. The true absolute is not, and never will be, quantifiable.

Nevertheless, current science is adamantly adhering to the scientific methodology as the only acceptable approach to acquiring truth. This is only *quantitative*, that is, treats objects as made up of parts. The q*ualitative* essence within a body is not made up of parts—though the object in question may still have parts. The qualitative state may be underlying the parts, and consequently a purely quantitative analysis will have limited success.

A simple analogy here might be helpful. Imagine a geometric type of existence in which the non-quantifiable aspects or absolutes be represented by circles. Anything quantifiable (that which can be measured, detected or analysed) is represented by straight lines, say, like match sticks. We can see that match sticks can never 'fit' with circles. One could only use shorter matches to simulate a circle. Circles would be incomprehensible to methods of detection if they were structures made up of straight lines.

We are encountering here *non-quantifiables*; factors which can never be detected, evaluated or analysed (and retain their truth validity). This is what we mean by 'non-quantifiable'; properties that can never objectively be detected, evaluated or analysed. Examples would be the 'aliveness' characteristic, the experiential and sentience attributes (basic consciousness). These attributes, of course, are not separate from one another, they are all 'quantitative' attempts at describing the single state, Absolute, and are thus features of a greater whole.

The Absolute must be recognised and defined as beyond the structures of particles, waves and frequency patterns, space and time; if this were not the case, it would mean there is separation within its nature, immediately invoking objectivity. In this condition there couldn't be any Absolute, and it would have then degenerated to relative values. If there are no parts (as must be the case), then it would mean that when it, the Absolute, interfaces with relative factors, it is always whole. In other words, relative to elements of manifestation it is everywhere at once—it is fully present in every point. It must be infinitely nonlinear, but in material existence it translates into linear forms and events, space and time. We shall see that the configurations, hologram and fractal, are interface systems precisely for the purpose of translating from a nonlinear condition to a linear one.

This 'everywhere at once' may be a difficult concept when first encountering it. Let us present a simple analogy, which may help.

Imagine large quantities of artist paint of different colours. We are not interested in colour itself. Let the colour represent a characteristic (of which there would be virtually an infinite number in reality). We now mix them together perfectly and spread the paint uniformly everywhere. As any artist knows, the final result will be a brownish colour. We again are not interested in brown, just the uniformity, but inherent within it are all the colours. We might ask, where is the red? Clearly it is everywhere at once. Imagine a spot of red paint outside this (say, left behind). This can represent a particle. So where is the particle after it is returned to the mass of brown (unified field)? It might be considered to be everywhere at once in a unified field. This is what quantum physicists are encountering when they detect and measure particles and then deduce, as per quantum physics, where the particle is when it is not measured (seen). There are comments like, it can be anywhere, in more than one place at a time; or it is like there is one big particle in the background. This background state is the quantum realm of infinite possibilities.

Thus we have a rough idea of what quantum physicists mean when they say the particle is in a state of superposition. Its position can be considered everywhere, and therefore its position is 'super'. Even more important we can now see that this quantum field (of paint) has all its characteristics in a nonlinear state. We now imagine projecting out tiny spots of paint; maybe a spot of red next to a blue spot. These two spots are now in a linear relationship, side by side and not superimposed.* These represent particles that can interact with one another and produce new characteristics.

[* Note the word 'superimposed' (layers going inwards—on top of one another) as different from 'superposed' (parts side by side). Even though quantum physics is the most advanced science on the planet it hasn't yet expanded into the inner space (a fractal system of nested levels of increasing degrees of order, which is what this book is all about—see later sections).]

Everything that is quantitative, that is, every aspect of manifestation that we call real, must *logically* have a reference—a basic context; a true zero. All our systems, including science at its best, are based on relative zeroes. This has a similarity to the difficulties with the covariant aether. A difficulty arose when the aether was proposed to explain the medium in which electromagnetic waves were transmitted—similar to water waves requiring the ocean. However, the state of motion of the aether couldn't be detected but, in particular, scientists, including Einstein, couldn't figure out how to formulate an aether theory that was covariant, meaning appeared the same relative to all bodies (moving). As a result, the aether notion was dropped (the special theory of relativity and the apparent constancy of the velocity of light required this removal of the aether)—see Section 21. This was a disastrous turn for the worse.

Thus all our external-world knowledge systems as viewed by science are essentially linear in space and time—they don't have the hidden context, which is like saying the twigs on the fractal tree are isolated (cut off). Consequently one will never find a basic reference which is a true zero (in linearity). An underlying non-linear condition is necessary. This would be the system of twigs attached to branches, etc. all the way to the tree trunk. Each fractal level references the next higher one. Finally we imagine the single state, the trunk, to be the Absolute—a final true zero, giving the phenomenon of 'everything everywhere at once'. It is a true zero; the perfect 'aether' that not only passes this particular test of truth, covariance (that is, it must look the same to all moving bodies), but in fact passes all tests of truth in physics. Everything, every feature of the cosmos, can relate to this immutable and stable condition equally. This does not require proof, merely understanding and the natural application of logic. Every separate and therefore different point in the manifest world is the same 'point' in the Absolute. We may see now that in going from the Absolute all the way to maximum separation in 3D, we go from a superimposed (nonlinear) condition gradually to a linear spread in space with particles side by side. We shall see later the significance of this in forming the basis of a fractal system of degrees of order.

We shall also recognise the term 'Absolute', as synonymous with basic consciousness, which is in an unformatted state that consequently must be an unbroken wholeness. It can't be anything but experiential, which we have indicated is non-quantifiable; and there can be no objectivity at that level. Note that the Absolute is the *now*, and is a true CAUSE (the point of Creation); anything that consists of particles and is relative, acts only as an EFFECT in the *now* (for example: in playing a musical instrument, the technique—based on existing programs—is an EFFECT in the now, but the musical expression is a CAUSE in the *now* (similar to intuition)—not based on recordings (study the cybernetic loop later)).

The higher the integration in the fractal system, the greater the unity as expressed by the carrier wave. But we must see the greater carrier waves as extending not just in space but also time. It simply wouldn't be logical not to have expansion in time when we consider the function of integration. In quantum physics, the tiny quantum, a whole energy, such as a wavelength (sine wave is whole), the quantum is at least 4D, it extends a tiny amount in time (its spatial diameter as temporal diameter). But quantum states can potentially extend up to any degree of wholeness, such as a universe or more. We are now picturing a gigantic wave in space and time. Science would call it a scalar wave—a potential energy wave, longitudinal with the electromagnetic waves inherent within it. The larger it becomes the more its effect is everywhere at once—remember it waves in time. This is demonstrated by shaking a rope, say, tied to a tree and creating waves moving along its length across space (that is, spatial waves) and taking *time* as with electromagnetic waves, but in this case of the largest waves, it spans the distance and just waves in time, one big wave, up and down (standing wave). See Appendix E.

One may see that as the wave spans more and more space and time it is becoming closer to representing the Absolute. However, the big waves carry smaller ones; it can't be the other way round; again telling us that the whole comes before the part.

As we have seen from previous sections, particularly Section 5, our lives involve relative zeros. We could even imagine a fractal scale of zeros. Logic tells us that there must be a final zero, an ultimate reference for everything else. Everything that is quantitative is contextual (based in references). Thus we should at least postulate an Absolute (reference). In 3D a lower reference will have its context in a higher reference, then again on the next level, but ultimately reach the final zero, that ultimate stillness that one hears about in meditation practices, religious philosophy, etc.

The New Science

Lower fractal levels cannot detect higher ones. For example, our level of consciousness, by using only its own level of mechanics or frequency spectra, can't access the next degree of order, such as the soul level. That is, the human cannot know about the higher fractal levels externally (meaning using the universe *objective* fractal system of energies and frequencies). It must go inwards, within, inner space, to go higher, since this is the personal fractal system, which also, as with the universe goes back to the Absolute (Figure 6). Let us present here a few examples illustrating some of the characteristics of the Absolute.

Imagine a sphere of empty (quantitately) space in front of one, say, about 3 feet in diameter. Inside the sphere is the Absolute condition—for this analogy and demonstration it must not consist of matter, energy, space, and time. Outside the sphere's imaginary boundary, on each side is our right and left. Further, we have a top and a bottom, and a front and back. However, as soon as we move inside the sphere all these defined locations, such as 'right and left', are non-existent: no right or left, up or down, front or back. Every point inside is everywhere else inside.

This condition would also be quantum physics' quantum realm of infinite possibilities and potential. It intrinsically passes all the tests of truth or logic of scientific protocol. One would expect that any manifestations emanating from this Absolute would be holographic in nature; complete interconnection of parts from the whole to the smallest part.

Another very different example of the absolute is to imagine an uneducated primitive man who automatically assumes that the reason an object falls to the ground is simply because that is the direction of 'down'. This is an obvious assumption and a failure in tests of truth with a gross 'preferential format,' since 'down' is only relative to a more basic context. This direction may be 'up' for a different context, such as being on the other side of the Earth.

A third more obvious example is when someone responds to the term 'Absolute' by stating that it would be eternal, but then another person interprets this to mean it goes on forever. Thus strictly this is not correct, since 'forever' is referencing 'time', and there is no time in the Absolute. We continually place references or contexts in the wrong order. Just as a twig on a tree is referencing, or is in the context of, or is relative to the next branch, of which the relativeness continues up the branch system to the source, the trunk, our 3D is in the context of 4D, which is in the context of 5D, etc. (in an internal nonlinear higher-dimensional direction), and so on until the decreasing relativeness reaches zero at the Absolute (at the 'top'). See previous sections.

In summary, we have indicated the various relevant features, such as the wholeness comes before the part, the nature of fractals and the hologram, Holiness as in Christianity, and the bizarre observations of quantum physics regarding the existence of the particle and that when not detected it can be considered as being everywhere at once.[2] Furthermore, the tests of truth in science all indicate the higher truths are in the direction of the Absolute. To be Absolute there can be no parts (such as stuck together by forces), no particles, wave/frequency patterns, space and time, since these would be in the relative class. It follows then automatically that in the quantum realm science is coming up against the Absolute characteristics, the 'everything is everywhere at once phenomenon'.

The essence of the Absolute is actually totally subjective but this is universal subjectivity, as indicated within Quantum physics about 80 years ago regarding the experimental-method limitation. The word 'relative' applies subjectively but in the experimental method this is personal subjectivity. This was also dealt with in Section 3 on scientific limitations. Some further interesting features of the Absolute are the following.

The Absolute, which as we have seen is totally subjective, a single 'whole' mind condition and is the essence of being totally conscious, tells us then that objectivity equates with unconsciousness. The Absolute had to make part of itself unconscious by counteracting itself, causing separation and a not-knowing condition between those parts. This is then the basis of all universal structure and environments (meaning something to be separate from and explore). The Absolute would be unconditional and totally subjective (no objective separation); it would be pure knowingness but unqualified, regarding its infinite potential and it would be in a perpetual process of actualisation from subjective to objective. The Trinity (of religion) or primary triad principle (of the New Science) would achieve, '... free will liberation from the bonds of Infinity and fetters of Eternity, giving a three-fold Deity personalisation.' [3]

The fractal system provides us with relative truths from low orders to high. This means a relatively low understanding or perception of a life-form's existence can reach positive practical conclusions for knowledge and technology. That is, an apparently complete system to exist in. However, the 'success' of this can be a trap since the ego readily makes this knowledge complete and resists any extension of it. It is vital that any fractal level of objective knowledge is recognised as being part of a greater whole (context) to be expanded into at a suitable time.

Behind every separation there is a fractal degree of wholeness. An objective existence only made up of particles stuck together is not practical and can't exist. There must logically be internal interconnectedness, which would include the unification of all opposites in the Absolute. The most important logical guidance in science is in the tests of truth of physics.

Notes
1. Book: *Man, God and the Universe* by I. K. Taimni, Chapter 1.
2. Book: *Space-Time and Beyond.* Bob Toben.
3. *The Urantia Book.* Transmissions from spiritual entities.

7.

THE INFINITE FRACTAL

An infinite background matrix from subjectivity to objectivity giving a gradient of different degrees of order for the formation of fractals.

The subject of fractals is probably one of the greatest discoveries in science; everything obeys the laws of fractals. However, science does not explain what fractals are for—why do they occur?

The mainstream definition of fractals is that they are self-similar forms on different scales. The leaves of a fern or the branches of a tree are typical examples. No part of the natural world appears free from them: ocean waves contain smaller waves; a coast line has irregularities of shape similar on all scales (viewed at different heights); a ploughed field has a clumpiness within a clumpiness; the foliages of a tree from leaf to whole tree reflect the branch hierarchy in its groups within groups; pebbles thrown on the ground organise fractally; even the stars in the sky visually reflect the fractal organisation; and we might add to this list—though not recognised in current science—the 'clumpiness' of the universe (which the Big Bang theory fails to explain), such as planets within star systems, within larger systems, within the galaxies, within galaxy clusters, within super-clusters of galaxies, within the universe. However, some scientists studying numbers of galaxies claim that the grouping ratios do not indicate that the universe is fractalised. But they are only looking in 3D. As we have already indicated, it is much more complicated than this. The whole system is fractalised going inwards into inner space; that is, in a 4D direction, as described in Section 1—what this book is all about.

All these fractals might be referred to as regular systems. However, fractals are inherent in all manifestations. Many systems of an irregular nature, such as turbulence of smoke or fluids from a faucet, or of a waterfall, also exhibit fractals but these only seem to become apparent when the randomness reaches chaotic activity and we have 'chaotic' fractals. Some of these systems can be studied using the mathematical tool 'attractor' which gives a graphic demonstration of regular and irregular patterns and, in particular, chaotic systems. However, even in the latter case the behaviour is not merely random; organisation is present within the randomness but there is still unpredictability of its future state.

We do not need to dwell on these more cryptic fractals. We are interested in why fractals occur and what their origin is. Could we have a universe without fractals? What would happen? The universe would run down. Entropy would rapidly go to a maximum and complete disorder would ensue (as per the principal model of current orthodox science). It would be a universe of 'particles' scattered everywhere. Fractals are one of the keys to life and, as we shall see, reflect the multidimensional nature of consciousness. Furthermore, if universes and dimensions weren't ordered in a fractal series and we had only one universe level (fractal) the intelligent life forms in such a simple universe would range from infinite intelligence to very low intelligence, existing at the same time—an unworkable habitat.

Now we might consider there are two main groups of fractals: linear and nonlinear, though we shall see they are the same. For example, branches on a tree, from trunk through smaller branches to twigs, are perceived visually and linearly, that is, in external linear space-time—they are not superimposed. We see the branches as fractal levels. But these are structured from atoms which give us perception of a branch due to the fact that our perception and most scientific instruments respond to this spectrum of energies. We don't see the organising energy behind these fractals. The reason is that the energies exist in internal nonlinear space-time (with which we shall become more familiar).

Science hasn't recognised that all particles and entities throughout the natural world, for example, planets, star systems, galaxies, etc., are fractals and contain wholeness in their own right. The same type of spirals of higher-dimensional energy[1] for a tree, as it branches into smaller spirals (smaller branches), also forms the universe as its spiralling energy branches off into galaxies, which in turn branch off into star systems, then planets, etc. See Figure 4 (universal vortices). Leading quantum physicist Professor Bohm recognised these greater quantum states.[2]

Also not detected in current science, the joints of the arm exhibit these same fractal vortices—nature's computer system for regulating the angle of the joints in physical skills.[3] In fact the whole arm and its fractal joints demonstrate the basic structure of all things: hierarchical systems, fractals, quantum reduction (of information), chaos theory, and even how free will works. Note that the vortices are, in reality, spherical and should be visualised as encompassing, for example, the joints.

There would appear to be an infinite gradient of fractals ranging from maximum fragmentation to one whole 'quantum' state, containing infinite possibilities (and underlying this would be the Absolute*). This would be a plenum of potentiality, a matrix prior to material manifestation; or vice versa, coming down from the whole, one would go through increasing degrees of fragmentation or differentiation. Now, points selected on this gradient for manifestation would be the divisions between fractal levels, such as arm joints, or tree branches. [*The Absolute, or quantum reality, would contain infinite possibilities. Of these possibilities, life would create and select the most likely probabilities.]

The New Science

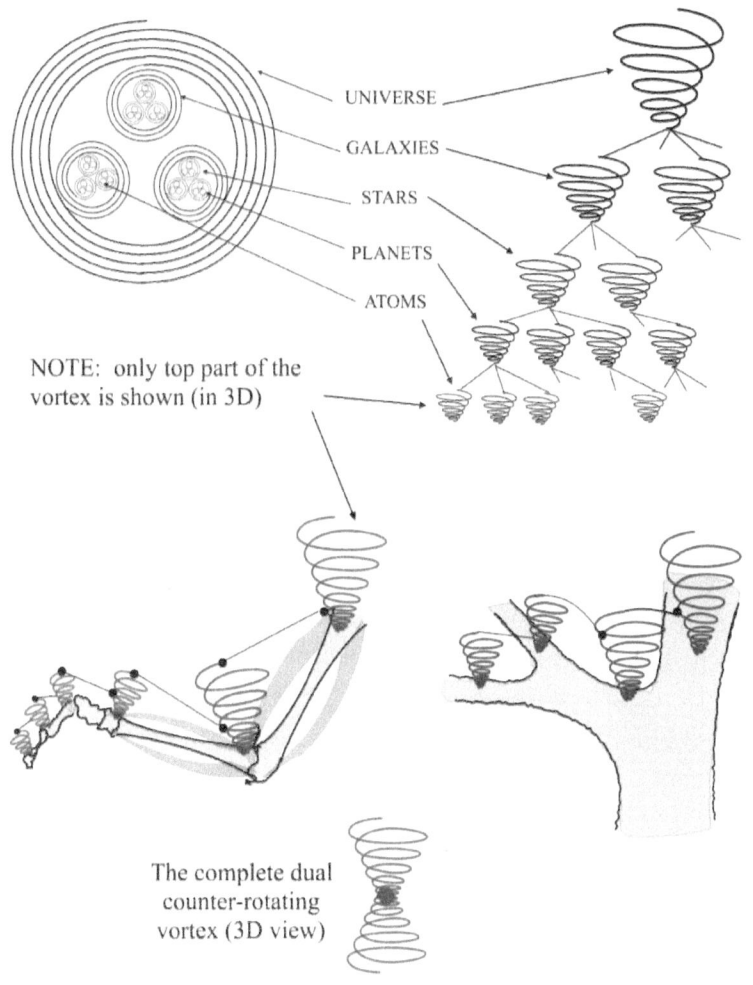

FIGURE 4 : The same fundamental vortex principles apply to all these fractal systems.

The planet Earth is more than the atoms which represent it in 3D. But we do not perceive these nonlinear higher-dimensional fractals in their basic state—if we could see them there would appear to be superimposition. Fractal levels are whole energy states in themselves, that is, a particular frequency, which are the vortex oscillations mentioned above.

The wholeness of each fractal level contains many probabilities—the probability manifesting in our third dimension is a particular selection. As explained in Section 2, the wholeness of all natural entities is greater than the sum of the parts. A galaxy is more than its stars, planets, atoms, etc. which *apparently* comprise it.

Thus underlying natural groupings of atoms, such as in forming molecules, cells, organs, leaves, planets, etc. is a singular packet of vibrations (the sine waves can be considered as large as each of these entities). The larger (higher-dimensional) sine waves control the overall geometry of the lower, smaller entities (and sine waves). These wholenesses on different dimensions are quantum states and constitute in effect a holographic structure, and the fractal levels are degrees of order originating from the hologram. See Figure 5.

From the holographic standpoint, we have the same configuration on different dimensional levels, but this is in inner-space towards the 'blueprint' level, and the outward appearance can manifest different probabilities. For example, with a more accurate understanding of the atom today, it doesn't look the same as a solar system, or a galaxy doesn't appear to have any resemblance to the DNA. However, underlying all these states are fractal holographic self-similar patterns—all manifestation follows these rules.

We have countless quantum states of different magnitudes impinging into 3D, keeping particles organised. There will basically be a complete gradient of contexts from the microcosm to the macrocosm but only selected levels in these strata of contexts manifest in the physical universe, such as a molecule, a cell, a planet, etc. (and of course universes are dimensional fractals within the multiverse). These are different degrees of order which have corresponding degrees of wholeness.

Does this explain the source of fractals? Not really. Why do these wholenesses (or degrees of order) occur? Remember we have stated that the specific manifested fractal states, such as a planet, are elicited from a continuous gradient and complete virtual state matrix of contexts. These are like the underlying fine steps of increasing integrity. Why does the presence of this potential gradient spectrum of contexts occur—the fractal gradient?

It is now particularly relevant to introduce the fact that perception manifests fractal organisation. For example, if one gazes at a clump of grass at certain distances away, the blades of grass will appear to group, mentally, more than they actually do physically. The attention is of course jumping around maybe erratically through different degrees of focuses but any groupings which exist in the environment, such as a whole house, causes the perception to focus into gestalt (whole) states. The intensity of attention is uniform across such a gestalt vision—it is one whole (frequency). All that is perceived is always in gestalt format whether it is small or large.

The mind is flickering about from one focus or state of wholeness of attention to another. These are quantum states, but they are in the mind. This applies to all experiences. There is actually no continuity along space and time (except at the Absolute level) but countless broken wholenesses, superimposing, of different magnitudes and frequencies. Why? What are these?

The same applies to the universe as the mind. The universe provides the external, objective manifestation, and the mind, the inner subjective experiential expression. Consciousness or basic awareness, whatever one wishes to call this, is beyond spacetime. If we begin with spacetime, this is not good physics. One must go beyond. That is, one must consider that consciousness primarily is not of space or time; it is an absolute. However, if we consider a universe is a manifestation from a non-space-time condition then the form it takes will mimic or reflect the nature of this non-space-time condition. A space-time interpretation of a non-space-time condition would be one in which energy exists only as wholeness of discrete states and will be capable of manifesting any focus, from the micro to the macro, and furthermore any and all of these quantum states could be superimposed. See Appendix D. An energy expression of a non-space-time condition would be holographic. This is the multidimensional nature of consciousness.[4]

This is the condition of our universe in which in itself it possesses wholeness, and within it, all its parts, arranged hierarchically, also possess wholeness down to a particle. These are all fractal states (manifesting from an inherent fractal hierarchy—the nature of consciousness) and are a natural product of a creative mind not bound by space or time. Moreover, the universe is part of a larger multiverse of which the constituent universes are arranged fractally.

Thus this fractal description of the universe is actually a description of the nature (structure) of consciousness, beyond space-time in itself but manifesting space-time from a myriad whole quantum states of every imaginable size, superimposing, giving the illusion of continuousness. If we attempt a direct description of the basic awareness underlying the energy of consciousness, we would have to say in 'selecting' a region of this fundamental awareness beyond space-time that this region must be pictured as having neither up nor down, left or right, front or back—this is the true nature of unity; it is an absolute. That means every part, or imaginary part, or sub-whole is everywhere at once. Clearly it would be infinitely holographic and nonlinear. Our description above is nevertheless a linear, space-time description (quantum states within quantum states, etc.) of this infinite nonlinear fundamental state of consciousness.

The fractal phenomenon is a system of degrees of order of basic intelligence and evolution, and also is the basis of the concept of Holy or Holiness (wholeness) of religion. When applied to multiuniverse systems, fractals provides levels of existence from high order to low (such as our 3D) for the exploration of the infinite possibilities of the Absolute or quantum realm, and these degrees of order have their basis in the hologram, which is a basic interface system between the Absolute and the relative.[5]

The New Science

We can now deduce that an infinite fractal forms a background matrix with a fine gradient of different degrees of order within which selected levels of fractals are fixed within a dimensional frequency scale. Thus this is a by-product of the Absolute consisting of an energy gradient, presenting different degrees of order within a dimensional scale, starting from high order to low. Compare the fractal system of a tree in similarly going from the highest order, the trunk, to the branches of decreasing degrees of order to the twigs (bottom of the pyramid configuration). For the fine gradient we would have to imagine an infinite number of twig/branch sizes from the smallest through to the trunk. Or use the analogy of a company organisation with its pyramidal ranking system with the president at the top (compare tree trunk). For the fine gradient the lowest level of ground-floor workers would have to consist of infinitely small units, and as we move up the gradient there would be a two to one ratio (say, two ground floor workers to one supervisor), and so on up the ranks. We can also use the human arm as a demonstration of fractals. For the fine gradient we would have to imagine an infinite number of joints from fingers to shoulder. See Figure 4.

As already indicated, the purpose is to form steps of increasing dimensions and frequencies from low evolutionary intelligence (degree of order) to the highest levels, and finally all information from these explorations are fed into the unqualified (neutral) Absolute, as it qualifies and formats itself into 'personality' and beingness formations. There will always be only one Absolute, but formations of magnitude, almost infinite, such as the Trinity of almighty entities (in the religious experiential/subjective interpretation). We shall return to this latter aspect from the science point of view. Thus the Absolute will have both states unqualified and qualified. The latter would contain formatting, externality and objectivity.

Why does the particular structure of fractals occur? See Figure 5. This is roughly a fractal distribution. The randomness is a fractal variable; meaning it could look more random (less obvious groupings) such as pebbles scattered on the ground, or blades of grass or groupings of wild flowers; or it could be more ordered, such as the distribution of stars in the sky. These are all fractal distributions. Note that in Figure 5 we have deliberately slightly

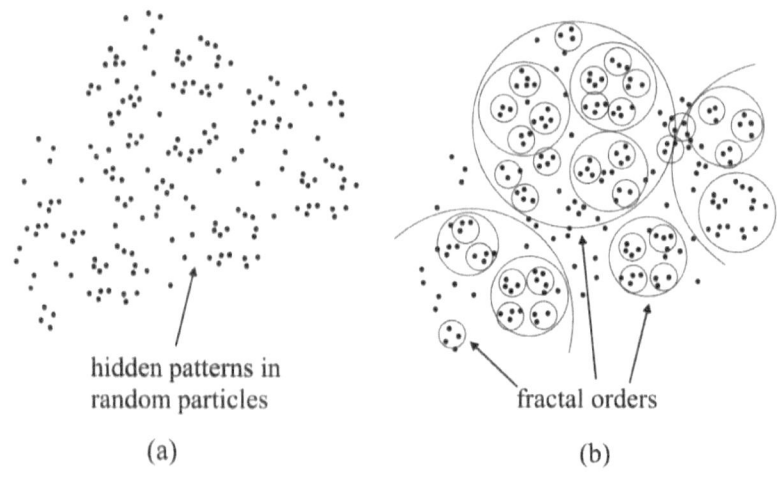

FIGURE 5: Fractal particle patterns indicated by circles.

biased the particle distribution so that the reader can spot the groups within groups.

Why are these special groupings occurring? This is a mystery in science. We need to recognise that underlying fractals is a fractal gradient. This is also a scale of frequency values or wavelengths. The sine wave representing a wave length is recognised in science as a whole unit of energy—a quantum state. It should be thought of as a pulse, turning on and off. As it comes into 3D it spirals in and spirals out. Now frequencies vary over the whole scale from low down into 4D, to infinitely higher when approaching the Absolute at the top. This occurs although the wavelength becomes shorter as the frequency and energy intensity of the wave or quantum pulses in space are greater (but note they become oscillations in time or higher dimensions). So we picture the higher-frequency quantum states as larger. This makes sense since in coherent structures the larger wave has to carry the smaller ones. To use the screen analogy, if we had a quantum pulse coming into 3D from the higher dimensions of the screen, this would thus give overall an image of countless spots of light of all sizes on the screen—picture discs of different sizes and colours; larger discs are higher in frequency and smaller ones of lower frequency. See Appendix D. All of creation springs from this infinite sea of superimposing energy waves (analogy: discs of light).

We now have the basis of why distribution of particles or matter appears organised but also possesses random characteristics; see Figure 6(a) for a guide. Underlying the 3D particle level are larger quantum states (fields) which keep the lower levels (such as particles) in order. For example, the particles may resonate as sub-harmonics of a higher frequency and larger quantum state. The latter acts as a morphogenetic field: a form-holding energy pattern, causing different degrees of groupings, for example, the clumpiness of a ploughed field. A smaller grouping is further down the scale than larger ones.

As a finer point or objection, one may still ask why is the 'clumpiness' (or organisation with organisation, or order within order) not smoothed out since the quantum states cover the complete gradient? Imagine the initial states of the organisation/disorganisation, such as the ploughed field of clumps of earth. The key now is resonance (of the oscillations of the quantum states). *Like* oscillations attract *like*. It's like a collective of people in a limited area who, say, range from good to bad. There will be a tendency to separate into two groups 'good' and 'bad' in spite of the gradient (degree of good or bad) between them—some random in-between groups will form; some with overlapping equal pulls. Thus we could scatter particles with a random action, but underlying the visible level (within inner space), if we imagine the 'spot-light' fractal quantum states superimposed, smaller groupings that tend to come together may be pulled into a larger quantum state by coherence. The larger state then holds the smaller group with some stability. The same will apply to all adjacent levels; see Figure 5.

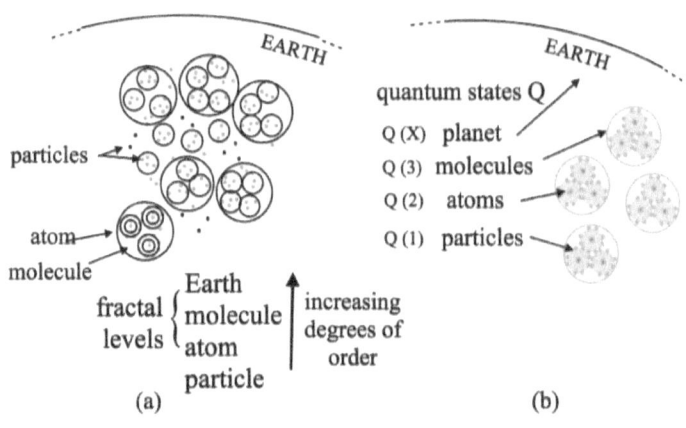

FIGURE 6: As per the New Science, particles, atoms, molecules, would have whole fractal energies. Higher up this scale would be planets, stars, etc.

In Figure 6(a) and (b) we see the formation of ordered fractals: atoms, molecules, planet, etc. Clearly, and certainly within the larger universe, there will be fractals within fractals. This level of fractals is as far as we need to go scientifically, that is, today, but we may well find not only fractals within fractals, but fractals orthogonal to fractals. That is, as we go up the fractal scale of dimensions into the multiverse of degrees of order, there is also a fractal scale branching off at right angles (90 degrees) with different characteristics of degree of order (analogy: columns and rows; go up the column to a higher 'fractal' level, or cut across along a row at any point, and this row will also give another type of increase in order of different characteristics).

Now since time is a by-product of matter, energy, and space, it will also be fractalised. This must be the case because higher fractals involve greater degrees of order, integration or wholeness of energy (quantum states). This is basically what we might call an infinite fractal. Consequently time will also manifest the characteristics of an infinite fractal. We can think of our objective, common time as consisting of tiny *nows* played off one after another. This is our basic time reference or objective context. In dreams, even day-dreaming, we will find that time scales can change; this is psychological time. When listening to music there is no musical understanding when hearing only the sound *now*. Our attention must span more time, connect together the past to the present and sense the future. This is not intellectual; this span is a whole unit of time—a quantum state. This wider span now passes along the objective time line but is unconsciously connecting it to the unity of the melody (higher dimensionally this is a single state). At a very high level of evolution the whole of one's 3D life could be spanned at once.

Thus we have what we can call a fractal gradient hierarchy of *nows* all superimposed. But we focus on the time with smallest steps giving the bottom-level objective clock time. Higher times will quantum reduce to lower ones when viewed with 3D perception processes. We are focussed (and selecting) in a linear reality.

This is a brief reference to time as an infinite fractal and can be studied further in other material by the author.[6]

Notes
1. www.nhbeyondduality.org.uk. Article: *The Basic Energy Unit*.
2. Book: D. Bohm. *Wholeness and the Implicate Order*.
3. www.nhbeyondduality.org.uk. Article: *Physical Mobility*.
4. Ibid. Article: *Theory of One*.
5. Book: *The Emerging New Science* and article: *New Science Theory of Creation*. www.nhbeyondduality.org.uk.
6. www.nhbeyondduality.org.uk. Article: *Time as an Infinite Fractal*.

8.

THE INTERFACE SYSTEM

We require an interface between the Absolute and the relative. This is the ultimate cybernetics system that science is omitting.

In general, there are many kinds of interface, both in nature and man's technologies, in particular, computing. In the computing field, an interface is a shared boundary across which two separate components of a computer system exchange information. This intermediary interface can be between hardware components or software components, or across both software and hardware; also between computer devices and humans. The keyboard of a computer is a simple interface device between the human and the computer. We thus have brain-computer or brain-machine interfaces.

In computers the encoding system may be different for different computers, such as in the IBM and a Mackintosh, and an interface program will be necessary to link them. In general, different frequency ranges are not compatible, and a high frequency may have to be stepped down to a lower one by means of an interface. In the case of the Absolute we are going from a non-quantifiable state to a quantifiable one. This is like endeavouring to make a perfect unity (single whole) compatible with a quantitative state of many particles, in particular, different numbers of particles—constituting different degrees of order. In this book we are obviously interested in the interface between life and machinery, or between the Absolute and the relative. (Note also there are interfaces between degrees of 'relative'—such as in the fractal scale.)

We previously explained the fractal gradient and its relevance to the New Science. This background matrix gives us the medium for recognising the interface system as we translate from the highest level, the Absolute, down through the gradient scale of reducing orders of universe systems. In simple terms we are going from the highest state of unity down to the most fragmented level; that is, from a wholeness (this is totally subjective, like a single mind) down to total objectivity, involving maximum separation for the theoretical mathematical model of infinite, infinitesimal mathematical points. See Appendix D. Between these two extremes there are different degrees of integration/wholeness. We may by now recognise similar properties to the hologram becoming evident.

Imagine the hologram (which is recorded in a holographic/photographic plate) of, say, a sphere (a football if one wishes to have a familiar object). We know that if we shine the laser through the plate we shall see behind it a 3D holographic image of the sphere. However, we also know that we can take any portion of the plate of any size (say, by cutting it into pieces), and an image of the whole sphere is obtained when the laser is shone through any piece, whatever the size. Thus the feature of interest is that all degrees of order from small spheres to one whole sphere are contained inherently within the holographic plate. (Recall that a degree of order is a degree of wholeness or unity.)

The fact that the hologram has all degrees of order in its images enables it to act as an interface between a greater order and a lower order—like splicing two disparate frequency ranges. Let us present an analogy for this mechanism. Picture a company organisation in pyramid form with the president at the top and below are the levels of the ranking system; say, executives, managers, ground-floor workers. Think of the president as holding all the knowledge of the company in a single gestalt of thought; this might be considered too high a level to communicate to the ground-floor workers. It would be more efficient and effective to step down his information, reformat it for the executives, who in turn reformat their degree of order to match the managers and so on. One can choose the degree of ranking gradient (how many ranks). In terms of physics this formatting would be the quantum reduction of wave patterns. In effect the president is the principal carrier wave for the whole ranking system. In turn, the executive level carries the managers, and the latter carry the ground-floor workers. This represents a fractal system (but the actual company organisation is only a simulation of a fractal system). The most fundamental interface system of which science is quite oblivious is occurring everywhere at all times—it is the essence of what keeps manifestation in continuous perpetual operation. An excellent example of the hologram interface is the learning pattern. The learning pattern is a 4D holographic template, which houses programs and converts nonlinear information into linear information. It connects frequency patterns containing programs with states of consciousness. Since this becomes also a subject of the mind we shall continue this in Part 2, in particular, the section on action concepts.

The relationship between man and the universe is given in Figure 7. The multiverse or cosmos has its own dimensional levels of frequencies, fractally ordered, providing degrees of order of objectivity, increasing downwards. The objectivity manifests the external environment, the separateness characteristic for life-forms to exist in and explore their level of consciousness, manifesting the infinite possibilities of the Absolute. The subjective side is an inversion of the objective but is complementary and has a corresponding inner-fractal system of degrees of consciousness, formatted to match the external worlds on the same level. At the top of the scale the two sides completely merge, and strictly they are only separated in the gradient by degree of unconsciousness (objectivity) and consciousness (subjectivity).

As we come down from the Absolute, or maximum level of consciousness, focussing on the subjective side of the fractal system,

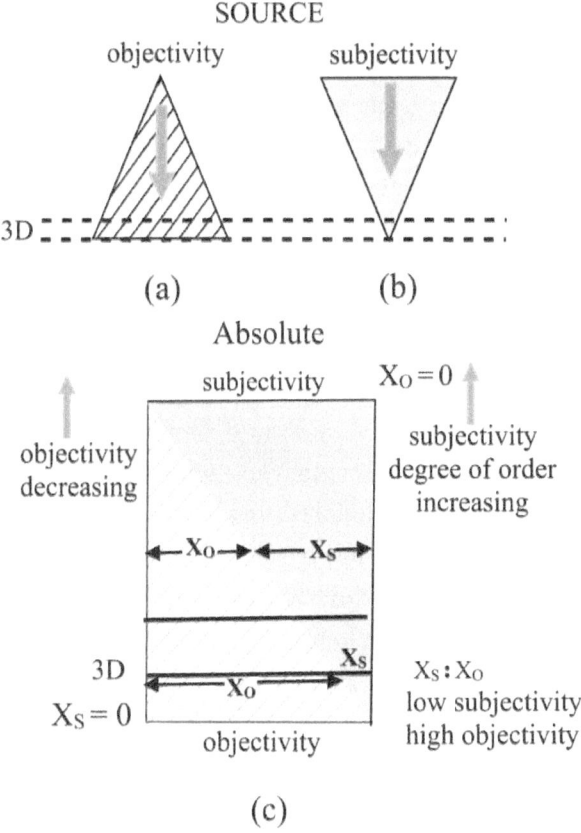

FIGURE 7: Ratio of subjectivity to objectivity as we go up the fractal scale.

in effect, consciousness is being reformatted all the way down this fractal ladder. Consciousness utilises an interface system at each dimensional level at which a universe is manifest, down to the interface of soul-level consciousness. This level then projects a fractal portion of its consciousness down into, and formatted by, the 3D framework/template (actually comes through the morphogenetic field of the planet; through the centre—higher-dimensional geometry is required). This is then formatted again by the 3D mind level imprints, and again by the brain, and finally our 3D body vehicle is a matching format for our environment. Note that this would mean that the soul aspects coming into the human body would contain all the upper levels of higher-order mind and associated cybernetic interfaces up to the Absolute.

All these transduction points are achieved by what we are calling a cybernetic process, as a *less* relative degree of order is translated into a *more* relative degree of order at each fractal division. Thus all these contextual interfaces have their own relative zero. However, at the highest level, the Absolute will be interfacing with the first relative state, which will be energy (now quantitative). We might consider this highest state the first unified field; it would relate in religion to the 'Holy Spirit'. Thus the most fundamental state will be the true zero of the Absolute, and then there will be the first translation into the highest quantitative state, the unified field. This is what science should be studying and researching: the ultimate cybernetics between the Absolute and a relative, since this is not really 'up there' but is in every point of existence. See Section 13.

The discussion of interface systems will resume in Part II, *Action Concepts*, when we bring in consciousness.

PART TWO

MIND AND CREATION PHYSICS

9.

FORMATTING SYSTEMS

In a nutshell, all of creation is manifested through a series of formatting processes, utilising templates, from the Absolute to the lowest dimensions.

We could say that at the most basic level a description of everything would be that the Infinite (Absolute) is applying 'restrictions', boundary conditions, in other words, formatting, to manifest limited, finite forms of *itself* and its own powers. Scientists have asked whether the universe is infinite. However, we may now know from good scientific theories (such as Einstein's general relativity) that the universe is finite. (A solution to general theory indicates that the universe is finite and can be likened to the 'surface' of a hypersphere; if we were to travel in an apparent straight line across space we would eventually go all the way round and return to our starting point.)[1]

Although numerous sources indicate that the universe is finite, this refers only to our external 3D. In higher dimensions we can take a direction at right angles to our 3D, that is, go 'inwards', 'within', and this is where we shall find infinity. (Infinity is not a number but we simulate it quantitatively by thinking in terms of large numbers.)

Thus existences are finite to the degree of their order (for example, 3D has its degree of finiteness). Recall the fractal hierarchy from earlier sections; infinity is beyond the 'top' of the fractal order—expressed, even in science, as containing the infinite possibilities of the quantum realm.

Now by *formatting* we mean any action which imposes a framework for energy-pattern compatibility. We are interested in geometric formatting and intelligence, and not algebraic such as in our computers.[2] Thus this is the same function as a template, a 'mould' or morphogenetic field. Even the formation of a definition formats ideas by putting in limitations and boundary conditions. Dimensions format space, time, and energy, dictating the arena for exploration of consciousness and its expansion. Any focus of consciousness is formatted consciousness—it is equivalent to a hypnotic trance, and it may or may not be held in place by a rigid framework of energy.

If science hadn't already been under the influence of mind-control dogmas, scientists would have proposed a general theory of everything in which everything begins with one whole (quantum state). It would, of course, be another alternative theory, just as we have Darwinian modes of origin, including the notion of building up forms from the minute. But beginning with one whole, it would have been the first acceptable scientific hypothesis of this kind.

Much of what we are referring to has been covered previously but here we wish to emphasise the important role formatting plays. The initial thought, which is consciousness, takes a specific form. This constitutes a template. We now consider that this consciousness (which is function) condenses into structures: consciousness filling out these structures. The analogy for this has already been covered. It is the example of comparing steam and consciousness (ignoring anything to do with temperature). We know that steam can condense into water, and water can freeze into ice. Thus we can imagine, say, an ice structure, hollow inside and patterned so that it acts like a template when the steam flows through it. The importance is to recognise the rigidity of structure which enforces its shape on the infinitely flexible energy of consciousness flowing into it. Note that this is basic consciousness, like an aetheric Absolute.

A more complex version would be to imagine a 3D structure—a system of geometric hollows and tubes. This could be the human body or brain. The purpose of such a structure is to cause consciousness to be formatted (focussed, limited, defined, etc.) in a manner to express appropriate information—in fact to create a reality, an existence. The structure will extract (mould, focus, distribute, absorb, elicit) from consciousness the appropriate energy patterns, frequencies, according to the structure that consciousness created in the first place (this is partly the role of the DNA and its unknown strands).[3]

A learning pattern in skills is such a structure, which when the input (consciousness) activates the learning pattern, the learning pattern selects a preset focus (program in the learning pattern) of energy from consciousness so that consciousness does not have to concentrate hard and struggle to make accurate movements as in learning a skill, such as endeavouring to find the correct positions of limbs, etc. The learning pattern guides consciousness like a moulded pattern, shaping the liquid poured into it.

Note that the DNA, in addition to being a vast reservoir of recordings and programs for organic growth, etc. is also a formatting system for the mind *and* consciousness. It wouldn't be so strange for off planet visitors to incarnate into the human DNA template at birth and thus be 'humanised' by it.[4]

Using the ocean analogy, we can imagine a complex structure within this ocean of consciousness to be a human personality—a pattern of encoded consciousness units—and at a distance, another somewhat different structure for another personality, etc. We can include all other forms, objects, atoms, etc. in this scenario.

We are talking about the nature of programming in the universal system, which is geometric and governed by the principal variables: frequency, rate of recreation (or transmission); phase angle of frequencies (spatial/temporal relationship of oscillations giving shape); dimension (magnitude and scope of the energies); and dimensional orientation of these energies and particles.

Of course, we are also implying that thought itself originated these templates, frameworks and structures. The creation of an idea immediately imposes on the Unified Field an encoded format which moulds the minute consciousness units into a pattern. The frequencies of this pattern will then continue to attract by resonance, similar patterns, sub-patterns, or any matching frequencies, giving more life to the idea.

Programming can be good or bad; survival or non-survival. It can also be very restrictive. A simple somewhat harmless example would be the extremely well-embedded program for face recognition. Psychologists have shown that when we look at a face, in effect, a filter (framework, template) is overlaid on the visual perception. This template is a general format of the human face—covering all 'normal' faces. Psychologists call them 'schemas', which can be applied to all thought, learning and cognitive processes. When we perceive a specific face we make the adjustments to the template to conform with that face. If the face fails to fit well into this format, the individual doing the perceiving feels uncomfortable—they become prejudiced. Everything we do, including all thoughts, act in this manner—they impose frameworks and limitations, which are programs on our thinking, acting, and creativity. In effect, the robot side of the human is copying our sense experiences (and thinking, etc.) every split second with the purpose of aiding survival and compatibility in a particular environment. This effect of replacing consciousness' creative efforts with recordings and therefore automatic responses to everything, increases with age. This becomes part of the dying process on this planet—for mutated bodies, such as our carbon body with over 95 percent of DNA missing (not 'junk', as science tells us).[5]

Continuing with this model of creation, consciousness is being formatted, defined, and qualified by structure, which is enforcing its pattern. And a pattern means information, knowledge, memory, programs. An atom is a structure condensed by consciousness according to certain rules, or a program, which in turn is a coded package of frequencies and consciousness units. The initial formatting is from or within the original Absolute (the ocean analogy with the sea at rest). Formatting of a template is achieved by particles and waves, and continues at each fractal level as we come down the dimensions; it acts upon the prior condition and forms the template.

We can imagine now an ego-structure developing around the personality, causing less consciousness to freely flow through the structure or system (the person). What happens now is that the consciousness begins to think it is the structure since it is being the same shape and frequency—the medium poured into the template becomes the template.

This is the state of man today—he thinks he is a brain and a body. Note also how this can apply to more subtle areas, such as when a learning pattern is activated we think we are the learning pattern and don't realise there is actually an interface between the learning-pattern structure and the input function. We identify with the learning pattern—actually think we are creating at that moment complex movements when in fact we are using the machinery to do it for us—we just follow it; let go.

Keeping in mind the ocean analogy, the next step is to envisage one of these structured personalities as becoming enlightened and breaking through the ego structure. The consciousness energy now flows freely and the individual realises he or she is not the structure but consciousness.

Now since the basic consciousness is everywhere, the person, while simultaneously retaining the information in the structure, that is, the personality characteristic and abilities, etc., can also be consciousness elsewhere; for example, combine with another person's viewpoint. Note that this gives a reality on the spiritual notion that God is everywhere and everything; it is all one thing—one type of energy (when first manifested) that is basically whole and everywhere. With this enlightenment and development, a scientist could take on the consciousness pattern of, say, a black-hole, that is, *being* it, in the distant cosmos, and grasp directly the energy relationships. For a sufficiently advanced race this would be normal.

The above model is very simplified but is excellent at handling the test-of-truth failure of physics regarding the apparent strange disparagement between the consciousness energy and the abrupt change to a rigid structure, such as the brain or body. In the ocean model we see more clearly that the dualism of consciousness (water) and structure (waves, etc.) is resolved.

We have to imagine a potential range of structures from the heavy 3D one, such as the physical body, through other structures for memory, programs, mind, and also aetheric, emotional, spiritual structures (bodies), etc. ranging up through dimensional structures of greater expansiveness, probabilities, more flexibility, greater dimensions as we return to the Infinite/Absolute form of the 'energy' of consciousness (capable of being any information). But anything less than the Infinite or Absolute is a focus, a framework or format of consciousness. Thus all forms of consciousness are similar to hypnotic states. We are hypnotised within the third dimension. A learning pattern hypnotises one's attention (or part of it) to focus energies in a certain way. A fixed belief system is a hypnotic trance-like condition, which occurs when the belief (recordings) are focussed on.

In the above reference to the structured personality becoming enlightened, how does this occur? If consciousness is trapped in the ego format, it is a closed system, and consciousness doesn't know anything else. This brings in the subject of open and closed systems and self-referencing systems.

See other sections on these topics.

The whole of human consciousness could be formatted into a particular belief system or, in fact, simply a strong ego structure, and since there is no spare consciousness (or connection to higher aspects) to step outside this closed and self-referencing system (similar to hypnosis), the human cannot know that it is dramatising this state of mind. Patterns could be programmed into human consciousness to, say, be cannibalistic or justify ruthlessness, virtually anything because the energy of consciousness has been shaped this way (geometric intelligence)—it doesn't know anything else. However, if one recognises that consciousness has higher-fractal levels, such as the soul, even though the soul is greatly blocked from the average human awareness, this 'external' consciousness, the soul, is capable of manifesting and merging with the human consciousness and, in effect, then aiding the human extension to step outside the programming (formatting, brainwashing).

Science and education teaches, however, that there is nothing beyond the human level of consciousness, no soul, no higher fractals, no God. This is the key to entrapment of the human race. That is, to develop and teach the factors that promote this condition, such as 1) encourage extended ego growth, which forms a closed system of consciousness; the human level cuts itself off from the soul fractal, and 2) discourage right-brain consciousness, with the subsequent impairment of intuition, imagination, creative abilities and spiritual maturity.

The nature of consciousness is that it is natively totally open. To have an existence, a reality, we must of course form stable and closed systems. The closure is only relative though. A 3D object has some permanence (closure) but internally it is open, via atoms, particles and zero-point energy of space. This means a 3D artificial object (such as a motor vehicle) does not have a whole quantum state which would be connected up to higher dimensions directly, however, its particular, atoms, molecules are natural entities and are under direct perpetual creation (as per the New Science). Education teaches closed systems in both technology and human behaviour; a condition leaving mankind vulnerable to enslavement. In doing so it is achieving this by not only teaching knowledge that man is only the material human but also developing those aspects of consciousness (the left brain) which strengthens this belief, and suppressing those aspects, such as right brain, which then severs the human consciousness from its greater (fractal) selves. The healthy state of consciousness, aware of its higher fractals—soul, over-soul, all the way to the One self, Source—can never be trapped; there is always a more expanded consciousness that is 'outside' the lower state.

The interesting peculiarity that arises from the nature of consciousness is that one can only know what the consciousness energy can be—or is being. One can't know what is outside the focus or framework unless the consciousness expands into it, the soul fractal, etc. One doesn't know when one's perception is prejudiced unless one steps outside it. The focus itself is achieved and held by resonance of frequencies within the template and if we change our frequencies, which is a change in focus, the third dimension, for instance, will change, just as one changes channel on the television.

Notes
1. A hypersphere is like adding an extra dimension to planet Earth. We have a finite world but we can continue in an apparently straight line indefinitely across the surface. This is 2D space curvature. The universe hypersphere interpretation has 3D curved space, which we can't visualise.
2. Book: *The Emerging New Science*.
3. Ibid. Original information on missing DNA from workshops on the Guardian material by A. Deane.
4. Book: *Engaging the Extraterrestrials: Forbidden History of ET Events, Programmes and Agendas*.
5. Op cit. *The Emerging New Science*.

10.

THE CYBERNETICS OF CREATIVE AND AUTOMATIC AWARENESS

Two applications of awareness in a kind of super-cybernetic relationship.

We can break down consciousness into two distinct modes. Let's call them: creative awareness and automatic awareness. (Note that we may use the terms 'awareness', 'consciousness', and 'attention' synonymously.) Creative awareness is primary, and automatic awareness is the by-product of the amazing cybernetics between the Absolute and relative systems (such as basic thought and mind computer recordings). We shall make this clear shortly.

As indicated earlier, orthodox science omits the validity of creative awareness. In fact consciousness is considered an illusion or just a by-product of the brain (and therefore inferior). As a result of doing this, science is relegating creative awareness to automatic awareness, but furthermore is excluding/removing the awareness aspect in automatic awareness.

What do we mean precisely by creative as opposed to automatic awareness? Both these states are occurring continuously, often simultaneously. For example, we may be carrying out a series of common chores around the house, which have become very automatic. At the same time, however, we may be lost in deep thought about something that is occupying our full creative attention (awareness), but the automatic actions continue and are using automatic awareness.

Another example is driving our car on a very familiar route, such as one taken to work everyday, and being distracted by some thought for a couple of minutes. Our learning patterns (which are the essence of skill) are capable of driving the car without the need for the creative awareness. Thus we may find that we were quite unaware of driving during this period but we handled everything perfectly 'on the automatic', including turns, right and left, etc. However, even in examples where there is no memory of the immediately prior experience, it is possible to cast back the full creative awareness (input function) reunite it with the memories and re-experience the automatic memory, but now using the creative awareness to do this.

What is particularly interesting is that it is possible with training to re-experience these unconscious periods by quickly recalling the memory traces that are left. Quite surprisingly we can find that we were perfectly aware of what we were doing during these 'unconscious' periods. Our creative awareness merges with the automatic awareness recordings. But most important we will find that we were obeying the program of our learning patterns at the automatic awareness level. That degree of awareness was not sufficient to 'take charge'—the machinery gave the orders. The robot side of us was in charge and we just supplied a necessary small degree of attention to keep the automatic mechanism running. There was complete control from our automatic awareness, and the creative awareness was focussed elsewhere. The instructions were totally dictated by the machinery of the automatic learning patterns/programs.

An excellent example of this is that this human automatic awareness is a similar level that the animal experiences continuously. The degree or volume of awareness of the animal is below the minimum level that can voluntarily accept or reject its programs, in particular, instincts. In other words, the programs, mechanisms, such as stimulus-response, control the awareness. This level of awareness is equivalent to the human's automatic awareness.

However, the animal's instincts would give efficient survival behaviour patterns sufficient for the animal's needs and protection. A good analogy here for illustrating this mechanism is the following. (Note that at the moment, we are discussing the general relationship of automatic and creative awareness, which includes perception and thinking, not just action concepts.)

Imagine a wooden board in which are cut various grooves, even complex patterns of grooves. This represents the mind in which recordings and learning are 'grooved' in. Recall that life and the universe function on geometric intelligence (Section 18), meaning here that a pattern of grooves *is* information. Next we imagine, resting just above this board, a sphere of water (something fluidic). Forget about gravity. The grooves are activated when the water rests on them and flows in them. The water represents consciousness. For a human there is a fair volume of water and even when it is forming and using the grooves there is spare water free from the grooves. Thus the creative awareness can take charge.

A problematic aspect of this is that the human consciousness has considerable unconscious regions from negative experiences and these can be pulled into the grooves, containing recordings of the experiences. The latter formed without the cognizance of the creative awareness, forming the basis of all problems: irrationality, the brainwashed condition, insanity, and general mental aberrations and illness. In the case of the animal, the volume of the water is small and is easily 'used up' in the grooves, which means the information of the grooves takes charge. In effect, for the automatic awareness, the grooves force their information patterns into the 'water'.

In Figure 8 we see that we are dealing with a universal cybernetics between awareness and machine/robot (spirit and body). We have creative awareness on the right and automatic awareness on the left which interact with one another as in a feed-back loop. Is this a case of two dissimilar components interfacing? Yes and no. This is the 'I/not-I', the quantum physics interpretation of the 'observer/observed'—to be explained. The right side, the consciousness, has a much higher order of consciousness/awareness. Its quantum state—a certain quantum of energy and whole 'portion' of the Absolute—is much greater than what we have on the left; that is, the two sides are dissimilar only in degree (wholeness), but similar in kind. The left side, the machinery/robot, such as the human body, cells, brain, and 3D aspects of the mind, are made up of atoms, cells, etc. These are all whole quantum states, but are much smaller than the individual personality on the right (an awareness unit). The smaller units only combine to form a whole (holographic) energy, not an 'absolute portion'.

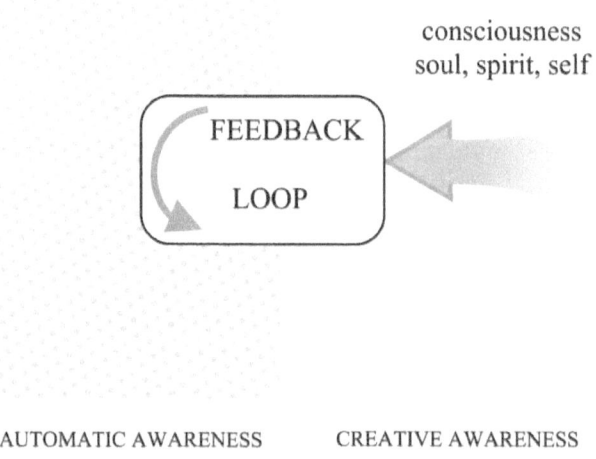

FIGURE 8

The robot side of Figure 8 copies everything the creative awareness experiences and then makes the information available to the creative awareness. Ideally this whole system, the feed-back loop of information, should be conscious with the automatic awareness relegated to perception in the margin of consciousness. However, as anyone may know, in the human, the awareness of activities in the body, brain and mind is very low. We are very unaware of the automatic mechanism 'assisting' us. This means the automatic awareness can often take over control of the creative side, which means controlling the individual's free will.

So how can these two conditions (left and right in the figure), interface? In Figure 9 the arbitrary numbers 1, 2, 3, 4, denote degree of order—just as we have explained with the fractal tree. Number 4 level could correspond to the twig level, and number 1, to the trunk level, or nearer the trunk. The frequency increases from 4 to 1 on a fine fractal gradient. These are degrees of order in sequence. This is an interface system (a little like splicing, as in DNA).

We can see now that the machinery of quantum states and frequencies can be made to match towards the left side of Figure 9, and the higher states of the creative consciousness (right) would attune to, or resonate with the greater orders, such as unit 1, which would be the largest quantum state and highest frequency. Also note that these quantum states are oscillations and that the frequency level is proportional to the size of the quantum state.

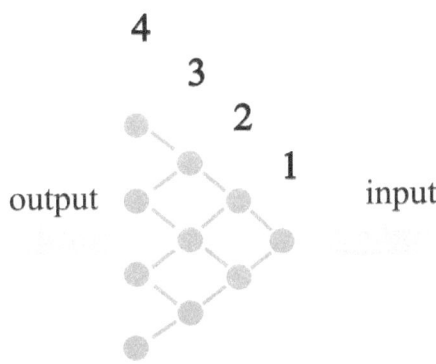

FIGURE 9: Simple representation of an interface mechanism.

The units in Figure 9 are oscillations similar to those in Figure 13, representing the learning pattern. Recall that the learning pattern is an interface system between the input/function/-consciousness and the program mechanism for producing results in our 3D environment. Thus there is a whole series of formatting systems down the fractal scale from a higher order to a lower one, that is, from a less *relative* to a more *relative* condition (and less *absolute*). This means a greater unity is interfacing or changing into less unity, that is, more fragmentation—hence the diagram in Figure 9, with a single state on one side of the array and a multiple one on the other. As we have stated previously, the single unit (at the top of the 'pyramid' shape) is a larger wave which carries those lower-order fragmented rows of units. A simple analogy is a company organisation, which we can view as pyramidal, with the president at the top, who represents the wholeness of the parts below him. What would be the best way for the president to interact with the many (say, 1000) ground-floor workers? This would be the use of intermediate-level members in the ranking system, such as executives and managers.

In Figure 9, this simple configuration holds the key to 1) the frequency step-down process, and 2) transduction from a state of higher order (unity) to a lower state, in other words, formatting. It is of course holographic. We could say that Creation itself is a cybernetics process. It starts from Cause and Control, then creates its Effect (which records) and feeds back this memory/copy to Cause, where Cause allows the Effect (to now be Cause-2) to aid Cause). The process is: Cause - Effect - Record - feedback - accept/reject. Regarding the 'reject', in more elaborate environments experienced in this way there may be undesirable events recorded (fear, pain, etc.), and Cause then automatically blocks (can't confront) the feedback of the recording coming from the Effect; the recordings are now stored unconsciously (out of the reach of Cause). See Figure 10 (more complete than Figure 8), which also indicates the (fractal) hierarchy of the cybernetic loop all the way back to the Absolute.[1]

New Science

We have not only the process of Creation here, but the basis of all problems—all it takes is to resist the unwanted Effect strongly enough to block off the recording. A mass of unconscious recordings may have accumulated of pain and trauma which, when triggered by a similar feature in a present time incident, will feed back to Cause this information (without Cause's permission or knowledge); in effect, the recordings/programs are telling Cause to get away from this present incident (experienced as a bad reaction) because it is dangerous (but of course it is only a *similar* thing contained in the incident). This feedback forces its imprint on the Cause (consciousness is like a flexible medium which can be moulded) which is then forced to re-create it (the past recording itself doesn't contain real emotional charge, just copies).

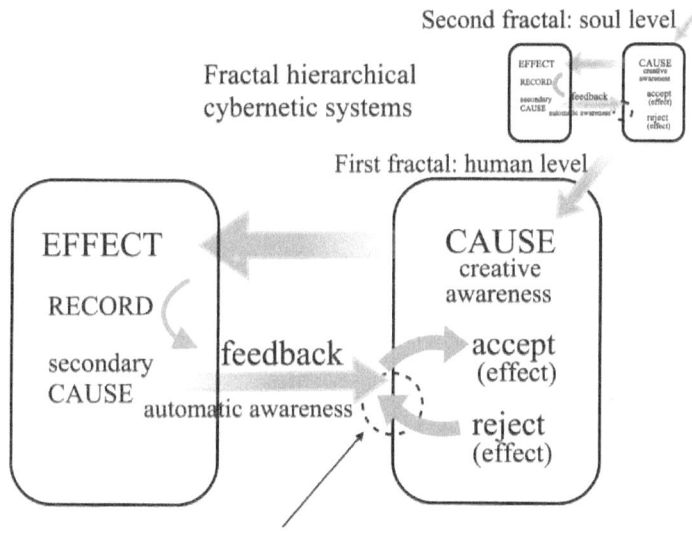

FIGURE 10: Ultimate cybernetics feedback interface system between fractal levels and also for all Creation and all problems.

This feedback loop is a microcosm of the macrocosm—same principle (holographic system). Thus the condition *reject* (deny, counteract, resist) in the loop, causes thought to objectify (reverse polarity) and become a separate (mental) mass in our mind. Compare a similar process at the macro level of the creation of the universe, or our environment, but at that higher level the 'deny' is done deliberately to objectify an environment to exist in and experience it.

Note that the feedback literally moulds our consciousness/-Cause (psychologists have called this a filter) according to all the data, recordings, it has accumulated. Thus the expression 'moulding the mind' is literal.

This same process applies to thought and the senses. Regarding the visual sense, the uneducated person thinks that we see by looking out. But science has determined that we obtain visual data entirely from light reflecting from objects. However, both are correct and act simultaneously, that is, 'looking' (attention energy going out) and 'feedback' from the light coming in. They blend perfectly and indistinguishably. This cybernetic process is much easier to observe with physical movements. The learning pattern feeds back the learned pattern/program—acts like a detailed mould/filter and shapes our attention/conscious focus into all the details for, say, a complex piano sequence. We think we are doing it consciously but we are only switching it on and off and guiding it, etc. There is more on this learning pattern in the next section.

Returning to the learning pattern, we have a good example of the relationship between automatic and creative awareness. When something is being learned, say, with fingers on a keyboard, or, whatever, the creative awareness must be used deliberately to move and control movements into the correct position and spaces. It is interesting to note that the reaction time for this creative (unlearned) control is relatively slow, about one quarter of a second.[2]

Now with sufficient practice the movements are learned, a program has been developed, and now the learning pattern is allowed (by the creative awareness) to control the creative awareness; cause and effect has switched round. Nevertheless there is a continuous automatic awareness accompanying the movements—but as stated previously, this level of awareness obeys the program (imprint, 'grooves' in the board).

With this state of automaticity the individual can place his or her main creative awareness on, say, musical expression, or simply think of something else, and the automatic awareness will keep the motion going—but all directional control is produced by the program. Note that in this state, the automatic awareness is 'embedded' in the machinery, and in complete attunement, which means the reaction time now is electronic to instantaneous. This is the reaction time in learned sequences. These two reaction times, creative, about a quarter of a second, and automatic, instantaneous, appear not to have yet been recognised by science.

In physical movements, this automatic awareness manifests as the kinaesthetic sense. Science evaluates this sense as something physiological. It is infinitely more complex, containing within it the whole hierarchy of holographic levels within the learning pattern, ranging from the smallest computer bit-type individual sensation to increasing degrees of wholeness capability of the learning pattern. On the machinery side of the learning pattern we have the diagrams, such as Figure 11, and on the automatic awareness side we have the kinaesthetic sense, which can be described in terms of action concepts in Section 11.

Thus there are two types of awareness or consciousness manifestations of function: 1) creative awareness and 2) automatic awareness. The first is highly aware and capable of developing and changing structure. The second occurs when a part of function is engulfed by structure, that is, the information/program in structure and obeys it. In this case, if the 'volume' of (external) consciousness is low, such as in lowly creatures, there will only be an automatic awareness with structure in control; or in the human example, the structures are sufficiently powerful and embedded to dominate the thinking and behaviour of the individual without the individual being aware of this; in effect structure is telling function what to do or think, or even how something is perceived. We can see again that structure relates to the quantitative (that which is built up and made of parts), and function to qualitative (single quantum states).

Notes
1. For a more complete description of this cybernetic process see *The Cybernetics of Creation*, www.nhbeyondduality.org.uk.
2. www.nhbeyondduality.org.uk. Article: *Reaction Times: The Two that Science Omitted*, and book: *The Attainment of Superior Physical Abilities*.

11.

ACTION CONCEPTS AND THE LEARNING PATTERN

How does the 'idea', a single state, translate into a complex multi-bit pattern in its applications to, for example, physiological movements?

There may be initial difficulties understanding this mechanism since the action concept as defined here is not yet recognised in science. It is part of the interface system between 1) initial thought or idea, including the intention, and 2) the output action. There would be no such word as 'action concept' if we use the scientific model, consisting of no real function, just a body/brain/mind entity with no real independent control. The word 'action' refers to physical tangible operations and the word 'concept' refers to mental intangible operations. Action concept is that part of the attention energy that activates corresponding units of the machinery/programs, etc. The programs thus translate a single quantum of energy (idea) into bit patterns, consisting of billions of computer 'bits'. These two operations, *concept* and *action*, must be interfaced. There are many similar interface systems within the body, as indicated in the earlier section, such as all the senses, mental thought, and physical action (the latter was used for research on the action concept by the author). See Figure 11.

Thus we are applying the action concept to the function of executing physical movements, such as in skills. Think of a simple movement of, say, the hand, from position A to position B. Firstly

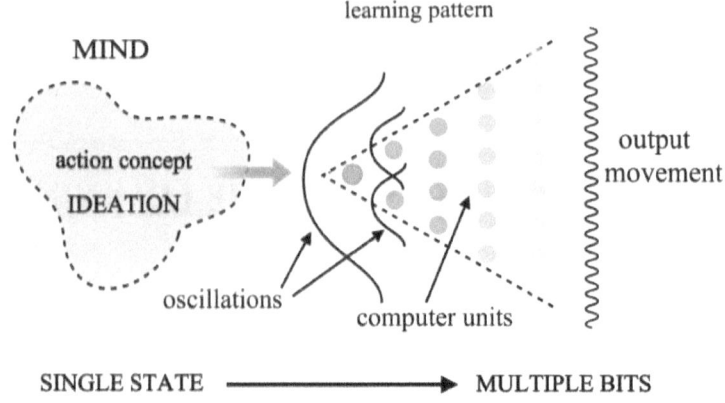

FIGURE 11: Interface between idea and mechanical movement.

the action is ideated (in the form of an idea). This is a single quantum or energy state, a frequency pattern, for this motion. The individual thinks the idea of the whole movement, in fact, essentially thinks of the end result. A mental image of the position of the hand at B is created. Note that this mental image at B may contain a visual picture of the hand at B but more important it contains a kinaesthetic perception, which gives a feeling of the hand at B.

The kinaesthetic sense reveals the programs for the movements. It is the interface between 1) the input consciousness, which is the ideation (and 'attention' energy), and 2) the program, which provides the details of this movement. This program recording is also the memory of the physical movement. The action concept and ideation, before making the movement, can be felt within the kinaesthetic sense, with or without an accompanying visual image. A single action concept is sufficient to control this movement (from A to B). But what if the movement of interest is not learned?

New Science

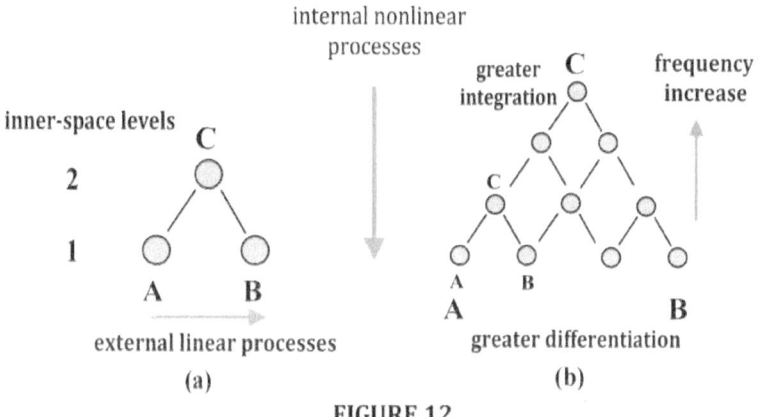

FIGURE 12

During the learning process we must focus on the parts and not the whole. The parts must be joined or coordinated. One automatically uses the largest, already linked, parts of the learning pattern that are learned and connects these together. It takes a certain amount of effort and concentration, in other words, deliberation, to link together movements (that are not already learned; this takes the longer reaction time of a quarter of a second, mentioned in the last section). The mind must envelop two separate motions as one whole. That is, by practising this, the learning pattern develops with a single action concept/thought that now controls those two movements.[1]

Figure 12 shows the coordination of movements A and B by means of C, which is recorded in a separate layer. Technically this (C level) is slightly higher, dimensionally; it will have a higher frequency. Thus now, with this configuration, A and B can operate independently at the lower level, or together, at the higher level. The coordination occurs on a different level, that is, nonlinearly, as distinct from linear coordination, which artificial computers/robotics would use. Note that this configuration is basic to all creation. In fact, and to digress a little, the triangle CAB will be the basic computer element for life's creation; and C will monitor the integration between A and B. This would be the equivalent of two fractal dimensional levels, 1) A and B, and 2) C. This is the triad principle again. If C is severed from A and B, consciousness at A and B contracts and chaos and negativity can set in at the level of A and B.

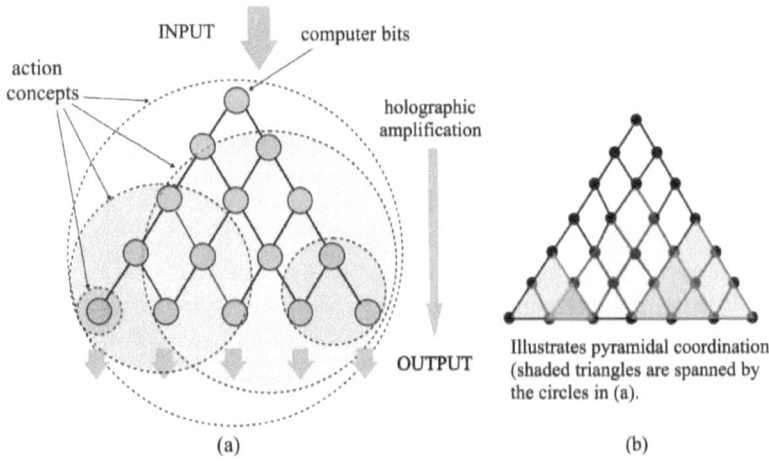

FIGURE 13: Holistic, holographic learning pattern

Now Figure 13 shows a larger learning-pattern diagram. Note that the circles represent the action concept; the attention energy of consciousness. The 'pyramids' in Figure 13(b) represent the structural aspect of the action concept. All the dots or bits can be connected to form a complete gradient of pyramids, as is the case when a combination of movements is thoroughly learned. But they can all be operated independently, ranging from a tiny motion (small pyramids) to a larger one (greater pyramid). This attention energy of the action concept is like a spot light that can instantly change focus, but flickers on and off—in reality these are quantum states.

All these pyramids represent action concepts. The mind or input consciousness ideated the action concept in the first place, but with some effort during the learning process. The movements are recorded and repeated. The learning pattern then feeds back this imprint or recording, which acts like a template to the input consciousness, and literally moulds it. The learning-pattern template, as it develops, draws from (absorbs/moulds) the input energy/consciousness and vaguely directs its focus, and aids the concentration of the awareness of the action concept. It amplifies and distributes this energy to create the correct movements. The learning pattern aids the focussing of the action concept.

For a learned movement then, or learned movements, every possible (in theory) action concept is inherent within the learning pattern. This enables the input to control any state of the co-ordinated motions. This is the key to being able to access the system instantaneously (instant computer access by consciousness). For example, if one is executing a continuous sequence of skilled movements, such as playing music on a keyboard, if these movements have been learned well, one can take one's attention off the actions but return this attention (input consciousness), and instantly, smoothly merge into the inherent action concepts of the learning pattern to control or change any part.

In the process of learning a movement we form the idea of the movement. This becomes an action concept as soon as an action is made. The required activity is now occurring in the learning pattern program. While the movement is occurring, the mind computer, at the rate of many times per second, records the action. Now when the action is repeated, the recording (which has formed), plays this back; in effect superimposes the recording of the first movement on the second to assist the second. Review the cybernetic steps in Figure 10.

The sequence: First we have the primary flow, creative consciousness, then 2) This flow creates the desired effect; records the activity of the primary cause, which may be perception detection of environmental aspects or operating a learning pattern in physical movements, and 3) the recording effects in (2) are fed back to (1). This energy ideally merges perfectly with the primary case such that cause (1) and effect (2) can be indistinguishable. Finally (1) performs the selection 'reject' or 'accept'. In the case of the learning pattern if one makes a mistake one does not accept it and the recording will not be embedded as intensely (hence the interesting observation in experimental psychology of the so-called 'learning during the interval of rest'—the errors fade more rapidly than the successes). If the movement is acceptable then the recording is allowed to build up.

Now when a person is performing a skill, the attention is continuously moving from one action concept to another; it doesn't have to but this is normal. Any focus will keep the activity going since the learning patterns are holographic and holistic (the input can span any of the circles in Figure 13(a) or triangles in Figure 13(b)). This is achieved by resonance. The frequencies of the input function/action concept are compatible and matched within the learning-pattern structure. A perfect interface has been created which actually causes the individual to 'dramatise' being the learning pattern, which becomes indistinguishable from self or consciousness. The person thinks he or she is controlling consciously the individual movements. The consciousness has in effect become the 'robot'. However, at any reasonable speed, the attention is focussed on the larger action concepts. Note that these levels or strata (going up the pyramid) are nested, that is, within one another. This is inner space—it doesn't go 'up' but *within*.

Each action concept of different size has a different frequency; the larger the size, the higher the frequency, but when the learning pattern is complete, that is, the movements have been learned, all the frequencies are in resonance. It is a holographic pattern extended in time, depending on the level of skill. This means the learning pattern acts as one whole. It spans a time interval—about one second for fast movements and high skill. Figure 12, remember, illustrates either an element of a larger learning pattern or can be regarded as the smallest unit regulating a smaller action (meaning, as we 'magnify' it we see the same pattern all the way down to basic computer bits, which are too small to illustrate).

This interrelationship between the parts, as already indicated, is both holistic and holographic; all internal frequencies within the learning pattern (when learned) are in resonance. Any one action concept activates the whole group (in a learned pattern). Also the power of the total information of the group (whole learning pattern) can be focussed into any part. This means any position is reinforced by all the other positions, before the positions have been occupied. They are preset when the intention occurs. This is a perfect system. This is holographic amplification.

Holistic systems, which science ignores, are difficult to understand since any one resonant part effects every other part and the whole. The energy is hugely amplified by this. The holistic interconnectedness between the parts, sub-parts and whole is not only in space but also in time. There is a holographic span in time of a degree governed by the development of the skill. During the execution of the movements of a learned skill the kinaesthetic sense of the physical action manifests not only the information/program or sense of the present position or motion, but also simultaneously the future positions within the holographic span. The awareness portion of the action concept can grasp/sense simultaneously all this information but this is at least in the margin of consciousness. Moreover, every element within this span, even future ones, aids the present one being executed.

There may be difficulty in believing that all the action concepts, small to large, in a holographic (learned) learning pattern, can be experienced as one action concept and all the individual parts sensed within this in the margin of consciousness. Clearly the conscious mind cannot observe this; it can only observe one state at a time. This experience is mainly unconscious but vaguely perceptible within the margin of consciousness.

Although an action concept can be ideated without physical movement, that is, it can be felt within the kinaesthetic sense in actual physical skilled movements, the action concept creates a tension state in the muscles corresponding to the span of this action concept. When a sequence of movements is being learned, small action concepts will be used giving deliberate selections of tensions.

At the other extreme of complete learning (the holographic continuum), one whole action concept is sufficient to perpetuate continuous motion* and this is felt as a general stiffness (less with greater skill) in the muscle group involved. No selection of individual tensions by means of individual action concepts is necessary. The larger (whole) action concept requires no more effort or thinking than a small action concept. [* This doesn't mean one learning pattern can span, say, the playing of a whole piece of music. The learning pattern scrolls in time, forming a holographic continuum. This is somewhat outside the scope of this book.[2]]

Thus the voluntary control over the tension is governed by the action concept. For a learned sequence one automatically operates the widest action concept (obviously the maximum level of skill). The focus of consciousness has only to apply a constant tension in the muscles (or fairly constant depending on the level of ability attained) to perpetuate all the detailed finely controlled movements. It is a totally robotic output, that is, programmed, but at the same time the input is totally in control every split second; this is the ultimate cybernetics.

As a final summary, the input consciousness initially acts as cause, which creates its effect within the learning pattern by building up structures. These structures are like moulds/templates and now format the input consciousness and the action concepts. The input is now more refined and acts as cause again over the structure, which improves and again feeds back its format to the input; and on and on, continuously refining the program, enabling energy to be distributed accurately to provide the desired movements and *idea* is converted into complex physical movements.

Notes

1. Strictly, the attention, which is a manifestation of the input energy of consciousness, oscillates between the two parts to be linked. The attention (energy) resonates and duplicates the frequencies of one side, then transfers to the other side, causing it (by entrainment) to resonate with these frequencies. This occurs repeatedly, making the link in the coordination stronger, until learned.

2. Book: *The Attainment of Superior Physical Abilities.*

12.

BODY, MIND AND CONSCIOUSNESS

The interface and interaction between body, mind and consciousness and that the brain is a formatting system for the mind.

In very basic terms we might consider the mind as an interface system between the Absolute (quantum realm of infinite possibilities) and the universe—and by 'universe' we mean the Absolute's material (mass, energy, space, time) manifestations or creations. In effect, the mind interface is a formatting system between the Absolute or basic consciousness and an (objectified) environment that is experienced by a life form. For humans, the mind is further formatted by the brain so that the mind can function and communicate to the physical body and subsequently the surroundings or environment.

Let us clarify the nature and function of consciousness, mind, and brain. Mind and consciousness are sometimes used synonymously—however, strictly, basic consciousness would be the same as the Absolute, beyond particle, wave, space and time, and contains the aliveness characteristic. Mind, however, does mean both basic consciousness and mind aspects, recordings, etc. structured out of particles, waves, space, and time. The mind comes before the brain and consciousness before the mind. The mind formats basic consciousness, and the brain formats mind so that the mind can communicate to the body and environment. The brain is an interface and step-down frequency system. The mind is like the harddrive and full memory of the computer, and the brain is the temporary memory and operating system (RAM memory), which brings in a portion of the memory from the harddrive; thus provides the transient information on the screen, etc.

Mind consists of particles and wave patterns in space and time. It is structure, forming a computer system consisting of recordings it makes of consciousness's experiences. It would appear that all perceptions and thoughts are recorded many times per second. It is so advanced that even emotions can be copied and recorded. But note that these are only copies—they are not sentient or experiential in themselves. When they interact with primary consciousness, that is, the Absolute, the experiences will be 'moulded', regenerated (programmed/formatted) by the imprint of the recordings/-programs, causing the Absolute to re-create, even emotionally, the dictates of the recordings. This is where cause and effect switches round and the recordings take the role of 'cause' (utilising the Absolute)—and consciousness takes the role of 'effect'.

Consciousness has higher aspects; these correspond to fractal levels. Figure 14, similar to Figure 7, shows how consciousness divides up amongst the dimensions from high order to low. The Absolute is the same on all levels ('everything is everywhere at once') and all recording structures are its own creations. However, the computer recording system can go out of the control of consciousness and become automatic, in particular, as function loses sight of what it is (what self is). We shall continue with this later.

Let us clarify the relationship between consciousness and the Absolute, and also consciousness and mind, of which the latter is a particular area of confusion for people. Sometimes when we refer to mind we are including consciousness. The interface between them is so precise that they easily form holistic states (together act like a single whole). Strictly, the mind is an energy structure (composed of particles and waves), and basic consciousness, separate or free from the mind, is the Absolute (no particles, waves, space or time); it is eternal.

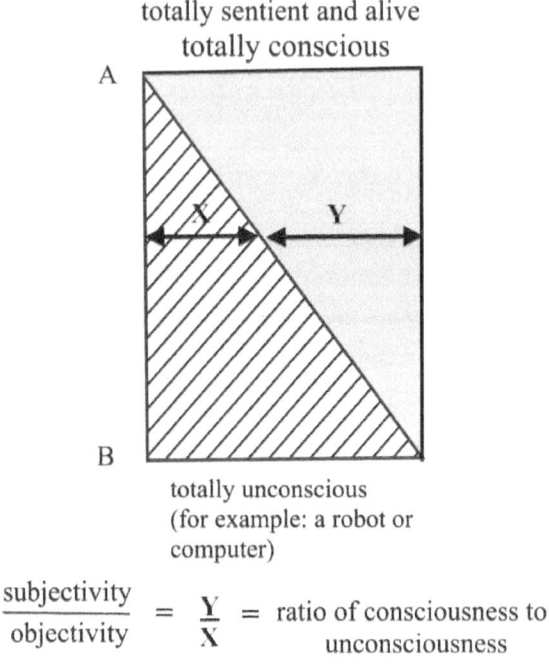

FIGURE 14: B is the theoretical lowest point in practice (but note that mere atoms have a degree of consciousness, that is, the Absolute).

Science and education teaches that the human is, in effect, just a brain and a body. The New Science shows that the human is infinitely more complex (see Figure 15). We can divide the human into two basic parts: structure and function. Note that science would agree with this, but unfortunately science interprets function as just another form of the same order as structure, and we don't get anywhere.

The structure refers to body, brain, and 3D mind energy patterns. Function is the input to this 3D human structure. (We shall see shortly that this input, mind and consciousness, is very complex.) In effect, then, the human can be regarded as 1) structure, an incredibly advanced robot, that science will not equal in a million years, and 2) a function, the main living entity, sometimes called the soul, which requires the robot for it (the soul) to function adequately in a particular environment. Let us try to increase the reader's understanding of this human configuration and the relationship or interface system between the 'robot' and the 'soul'.

A simple first analogy is to imagine the robot is a motor vehicle. We can now see quite clearly what the function is, which is the human in the driver's seat controlling the vehicle. Imagine our attention is outside the windscreen and we just think 'turn right' and it goes right, as we do with our body. However, now imagine introspecting one's attention inwards, behind the windscreen and identifying oneself as the driver, and that the reason the car goes right is because one turned the steering wheel for that direction. This is how the author obtained the information about the learning patterns of the mind computer system, that is, by introspection, to view behind the programs of the learning patterns. Note that science skips the driver altogether. Let us clarify this with another analogy.

Keep in mind the two components of the human 1) the robot, and 2) the input consciousness or controlling entity. Picture a computer on a desk with the human computer programmer seated at the desk. The computer corresponds to the robot and the programmer corresponds to the input function. It is clear that computer programmers (or computer scientists) created the computer in the first place; also that the programmer is in control and it is he or she who programs the computer; switches it on and off, etc. The programmer is *cause* and the computer is *effect*. However, when the program is complete the programmer allows the computer to play back the program to aid the programmer; that is, the computer now takes over *cause* and the programmer receives the *effect*. But the programmer is always in control.

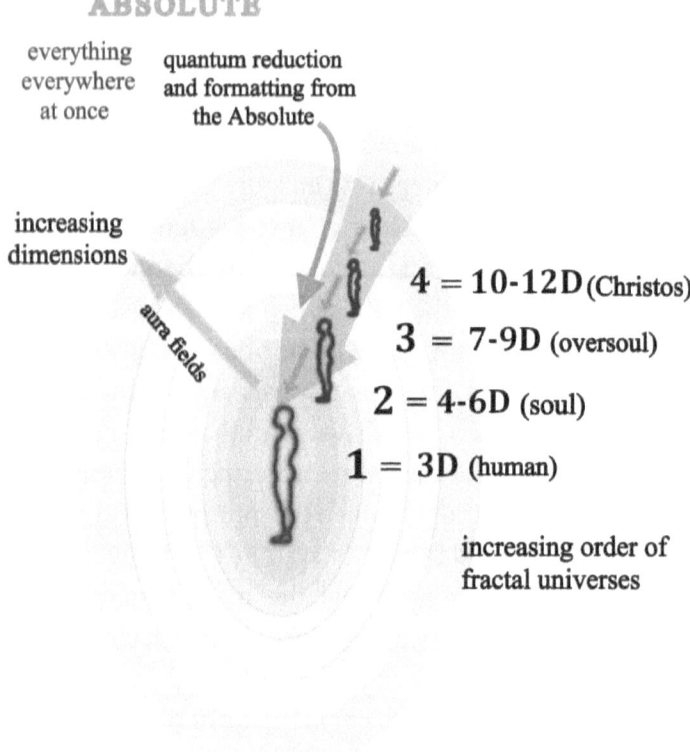

FIGURE 15: Repeated figures simply indicate currently existing higher lives in the inner-space dimensional universes.

The shocking news now is that science ignores and regards as an illusion the programmer seated at the desk (using the analogy). Science is desperately endeavouring the show that the computer (analogy) came into existence by trial and error (Big Bang, or from chaos, randomness) and, further, that the computer somehow learned to program itself.

One may think that this is a poor analogy since the programmer is obviously far superior to the computer. However, it isn't such an exaggeration as one may initially think. The 'soul', that is, the input energy or function, the conscious self, which is controlling the 'robot' is immensely complex. Using the analogy of the fractal tree this human soul extension corresponds only to the twig. The twig is part of a fractal system with higher aspects (fractal levels/orders), branches to tree trunk, that is, up to the Absolute (one whole).

Just as there is a holographic fractal system of universes, containing our external worlds and environments to be explored, there is also the internal ('within') fractalised higher aspects of consciousness with corresponding body vehicles, which match the energies of the environment and are in attunement with one another. The internal/external aspect is the subjective/objective relationship (which is not understood today in science and education, in spite of what quantum physics has revealed). However, the degree of apparent separation between the outer and the inner senses decreases as we go up the fractal scale of dimensions; everything is becoming more mind. Thus this is expressed by the subjective/objective relationship and precisely what quantum physics exposed some 80 years ago (known as the Copenhagen Interpretation and the collapse of the wave function).[1] See Figure 17.

Thus the human 3D consciousness, the function in the robot, is at the bottom of a long line of fractal, higher aspects of consciousness in repeated formatted steps from the Absolute all the way down to our 3D environment. Each of these steps is a different dimensional universe and furthermore each level of consciousness has its corresponding level of mind. That means at the 3D level, inherent within (and unconscious of) our 3D consciousness, are all the higher orders of mind (structure) up to the Absolute (which of course is a state prior to mind, since mind contains particles, waves, space and time—and mind is thus relative).

Keep these analogies for structure and function in mind as we take this a step further. As stated previously, the author trained himself in introspection of the attention/awareness, in effect, 'viewing' *behind* the learning patterns in physical movements. Science does not recognise that the kinaesthetic sense, which is sensed in our muscles and is the perception of position and movement, is in fact the interface between the input function, the consciousness of the individual, and the learning patterns. As a result of this understanding, we are then in a position to evaluate the nature of the program. The author discovered that the learning pattern was a 4D holographic template that houses programs and converts nonlinear information into linear information. Let us proceed with a further analogy, in particular, to make clear the purpose of the 'robot' aspect.

Imagine a robot sufficiently large so that a human can stand within it; that is, the legs and arms would have extensions and the body is completely enclosed. Even the human's eyes could have an interface system with the robot's eyes of extended abilities, such as, for example, x-ray or night vision. The purpose of this obviously would be to enable the human to explore an environment for which the human was unsuitable, such as on another planet. Next we imagine that the interface systems are so good that the human's attention eventually remains external; that is, he is no longer needing to pay attention to what he is doing inside the robot where the interface system is. We now imagine that so much time is spent in this mode of operation that ultimately the human forgets that he is inside the robot. In effect, he has forgotten who he is, and in fact now thinks his body is the robot. Or to make the analogy closer to truth, imagine that he is born in such a structure which grows with time and he has thus never known in that particular life anything different. This is the state of the human race today. People have forgotten who—or better still what—they are. (The author's other books and articles deal with how this has been brought about.)

Some experiential clues for the reader regarding the presence of this internal interface and where the real live viewpoint is originating, that is, not basically through the body/robot but via the input function or consciousness. Firstly obtain some familiarity with the kinaesthetic sense. This is not just the lower fractal of physiological coils in the joints, as science tells us, giving the sensation of movement, but extends into the mind learning patterns and programs. All one has to do is move, say, the arm, and observe. Even without looking at the arm one knows where it is. There appears a sensation in the muscles, giving the position of the arm and movement. Recall the human in the mechanical robot described above. This kinaesthetic sense corresponds to the sensation the human would experience as, say, his human arm manipulates the external bionic extension through the interface. The sensation is at the interface; one can obviously feel what the real arm is doing.

Another very simple example is simply recalling some memory, and internally, with the mind's eye, looking at the image. It is obvious that someone is looking at the picture; we are not the picture, which is a mind computer structure/imprint/recording.

Science is totally missing out on Creation's fantastic cybernetic systems. All knowledge and energy are contextual. The fractal scale of life and the universe shows that everything that is quantitative (energy, particles, waves), that is, can (or could be with more advanced instrumentation) be detected, evaluated, analysed, references something else; it is relative to another context. This is a gradient of relative values all the way up to the Absolute, which is the final true reference and stillness. The Absolute passes all the tests of truth in physics—though this may be arguable since it contains nothing quantitative and, in fact, is science's definition of 'nothing'.

The dimensional divisions are formed within this fractal, subjective/objective scale or gradient of frequency increments from low to higher orders, providing the basis for a hierarchical universe system. Each major context carries its own relative zero.

Now the intuition is a special ability, quite different from the left-brain intellect. It has the ability to access higher aspects of consciousness. The robot side of the human is context-dependent; the intellect alone can become stuck inside this context and, for example, fail to see the direction in which the knowledge is going. The intuition can bypass this restriction and has the capability, potentially, to contact the higher aspects of consciousness in the fractal hierarchy. This is a right-brain consciousness characteristic and not possessed by the left-brain intellect. Why are not people more aware of even just the next fractal level, the soul? We shall briefly see later the types of barriers that exist on this planet to prevent people from recognising the higher levels.

Notes
1. Figure 17 in the next section shows how the relationship varies between the subjective and the objective. Also see book: *The Emerging New Science* vol.1 for more information on the Copenhagen Interpretation.

13.

CREATION PHYSICS

Science is still in its infancy and has not yet awakened from our Dark ages to the glorious and magical physics of higher dimensions/fractals of the wonderous Cosmos.

In spite of many brilliant achievements of science and, in particular, the application of science in engineering, science can be regarded as still in its infancy. The main impediments are the negative ego, the programmed closed-mindedness of particularly many academics, and suppression of and control over knowledge, and finally the creation of artificial realities.

The foundations for such deviations from truth can ironically be found within the principles of creation and the nature of the subjective and objective relationship. Part of the problem is that all manifestation when coming into existence could be considered to be guided by a balanced principle between stability and instability. Stability requires counteraction of energies, which is the same basic mechanism as in the formation of problems (two opposing forces) and which also provides the stability for the formation of matter.

If this seems disagreeable to the religious as though limiting the power of God, consider that we exist in a universe designed in this way. This principle will test out a scope of relevant possibilities to be explored within the infinite. For example, stability can be overdone, slowing down ideal evolution (for a purpose?) and reciprocally, instability will encourage too rapid a change. We shall say more on this.

Now how does the system of many universes come about? How are the multiverse fractal dimensions formed? We have established a 'top down' theory (as are all religions and metaphysical subjects) and therefore it is an approach to creation in which the primary impulse will be *thought*; this is permissible with a 'top-down' creation theory.

If one is not happy with this postulate and assumption, we could use 'reverse engineering' to arrive at it.* We can begin with all the actualisations of existence; the history of manifestations, knowledge of our universe, galaxies, etc., and work back. That is, ask the question why is our 3D existence the way it is? To some degree we have already done this in early sections, such as on the whole and the part, and in volume 1.[1] We have seen that science does not distinguish between the whole (as a true undivided unity) and a 'whole' formed by parts stuck together by forces. We mentioned previously that there is not only a distinction between them but we could claim there is potentially an infinite difference. We have seen in volume 1 that Darwin's theory of evolution does not explain how life occurs spontaneously from organic material, or the gap between plants and animals, and also between the different species. Furthermore, the Big Bang theory suffers from a complete failure regarding the infinite regression test of truth in physics.

[* 'Reverse engineering' appears to have originated from devices which have been acquired complete and then their function evaluated by reverse steps of engineering concepts, such as in the case of reported captured UFOs.]

Now, at the level of creation there can only be consciousness, that is, there is no unconsciousness or objectivity (based on separation). This means the original Creation is a single state or Source; what we have called the Absolute (or the Unmanifest). See Figure 17. This is also the quantum realm of infinite possibilities (infinite wisdom). This was described in volume 1, but we can recap on this creation here using the triad principle to explain the mechanism of creation and manifestation. The Absolute becomes more qualified, which would naturally involve a degree of formatting of the infinite and bringing about a supreme entity. This would be a personalisation process, bringing the subsequent 'personal' relationship that the religious feels with God. This is not the basic unconditional, unreachable, unattainable Absolute.

In Section 6 we saw that the Absolute cannot be imagined except by using analogies and quantitative descriptions, otherwise we arrive at a nothingness. Let us recap on the nature of the Absolute. Think of a uniform, all pervasive aether type medium. This (the aether), however, would strictly be a homogenous energy layer (the Holy Spirit might be appropriate), and we must think of the Absolute as beyond this or underlying it, having no particles, waves, space or time—its whole is in every point; it is eternal, neutral, and unconditional. Everything springs from it. It is basically unqualified but its aliveness characteristic brings about a self-awareness and a desire to explore its infinite possibilities—of the 'quantum realm'. This very activity brings about a personalisation with no limitations (from the infinite possibilities); what could be referred to as first cause and beingness, Source, or God. Underlying this, the Absolute is unconditional and will allow anything—positive or negative, Light or Dark. However, since we are initially dealing with a single state of Beingness, any personalisation coming from the Absolute would never choose anything less than perfection, harmony, integration, ethics, since it is only dealing with itself; it would be totally illogical to harm itself or create anything destructive, since it is all that exists (the potential for the negative side can only come later as the separation occurs and free will is given to lesser parts).

At the level of the Absolute we subsequently deduce that there can only be *cause*. This means that unconsciousness must be introduced to create objectivity and existences, otherwise there can be no true effects or unknowns. At this Absolute level there is no *effect*. In order to have an effect, Creation must separate out a portion of itself; achieved by a degree of enacted 'denial' and the subsequent counteraction. This might be considered to be the key to the formation of the fractal hierarchy. It shouldn't be difficult for the human to understand, since we do the same, knowingly or not. Much of our processes of thinking and perception are mini-reflections of cosmic creation processes.

The counteraction is the opposite flow. In terms of consciousness this will make the projected portion into an unconscious state (an environment is an unconscious state for an existence). This counteraction immediately introduces the beginnings of elementary space and energy, and these two energy flows (the *cause* and receiving the *effect* of the cause) can be considered to have created elementary mass. It is as though there is a basic aether (an initial manifestation of the Absolute) that undergoes 'condensation' (somewhat analogous to water vapour condensing into water, and water freezing to ice). Thought would create basic templates in this manner by action/counteraction, from scalar standing-wave oscillations. This would be a specific form of a morphogenetic field (a form-holding pattern). Let us digress and explain a simple analogy regarding the initial approach of this Creation.

Creation occurs more in the manner of how a sculptor works—let's say in stone. He doesn't gather particles of stone and stick them together. He begins with a large mass of stone. Suppose he wishes to create a figurine. Inherent within the stone block are infinite possibilities. Note that he begins with the whole and pares it down—in effect formatting, limiting the result to selected shapes. He might pause at the stage where he has a shape, undefined in detail, which can be any human. This is analogous to a broad template or archetype. Similarly he continues chipping away at the stone, passing through the stage of selected male or female (more defined templates), and finally the desired specific human individual. These stages are like further templates or morphogenetic fields (energy/form holding). Recall the ocean analogy of creation.

The ocean represents more the unified field but 'underlying' this is the Absolute. We imagine in the analogy that we don't generally detect the ocean (seemingly empty space). However, if the ocean is disturbed; if there is motion and a shape (the two go together), this phenomenon now becomes detectable. Recall the example of Hertz passing radio waves from one room to another, but space was considered a vacuum, that is, the waves would be detected but not the mechanism/medium that transmitted them. To explain this, scientists postulated the aether (compare ocean).

Thus a shape in the ocean becomes a real wave, or particle (tiny whirlpool), or complex body in general, and ultimately the creation of a universe. Note that everything is merely a modulation of the 'ocean', furthermore, nothing actually transmits linearly (a wave passes on its disturbance; it does not bodily travel along the surface). We might note that particles do not actually have trajectories through space; they are merely wave patterns. There is apparent motion in the wave but it transmits its disturbance. It does not travel linearly. This means mass does not actually move (any more than neon lights, flickering on and off, travel in a pattern). The atomic oscillations, switching on and off, cause the atoms to appear in a new unit of space.

Returning to the formation of templates, one could imagine formatting the infinite down through increasing details, giving more general forms to more specific—for example, templates for a collective down to individual templates (DNA). The Creation plan progressively precipitates the fractal hierarchy. The primary state of Creation (Absolute) in our 3D terms would be that it appears as nothing, quantitatively, yet potentially is everything. But there cannot be anything outside it; this would contradict the logic of an Absolute state, or primary condition. When it creates divisions, resulting in elementary energy and subsequent objectivity, these are projected 'out'—created divisions—it does not divide in itself. This would be a contradiction since the two divided parts would have to be created from a larger context and the larger context would be lost; it is postulated as eternal anyway. The two parts have to reference a more primary state, which in this case, is of course the first state of Creation. We now use the triad principle to show

how a separate manifestation and therefore objectivity can occur for the process of exploring its infinite possibilities.

Note that in Figure 16, level-1 can monitor both sides in level-2, but if it focuses into A at level-2, it can only perceive B as external and separate. If the separation barrier was removed between the two levels, then A and B would immediately become C at level-1. However, at this early stage of Creation we would consider this to be flexible; there would only be *cause*, and any *effect* would be allowed. By 'flexible' we mean that at the limited level of A and B, both these can know their connection to C and go back to it under their own volition, and there is no actual and enforced barrier.

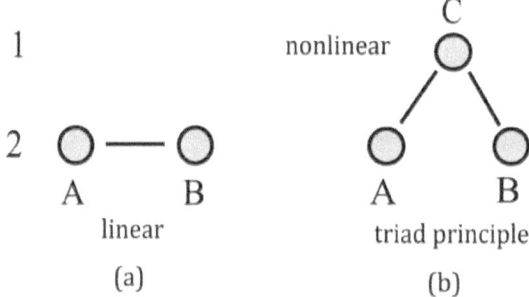

FIGURE 16: The figures show the key difference between science's linear robotic approach (a) to existence, and the nature of consciousness (b).

At the A and B level, A and B are separate. If the barrier exists (is a true 'blockage' but is allowed) then A doesn't know B, and B doesn't know A (is not aware of the internal connection with C). But A and B are C at the first level. Now imagine countless further divisions from level-2 downwards, with precisely the same principles in operation. This is the essence of the build-up of the holographic fractal hierarchy. Objectivity increases as the process multiplies away from the Source and it becomes easier to introduce a compulsive block/barrier between any level, preventing the lower order from being aware of the higher order or believing it exists. This is precisely what is occurring on our planet between the human level and the next fractal dimension up, which is the soul level. Unconsciousness increases as we go down the scale.

Thus the first division gives us the triad principle—which then continues to multiply as further divisions occur. This first step would form the religiously termed 'Trinity' and would give the first finite manifestation free from infinity, enabling the eventual creation of existences in space and time. It is important to emphasise that A and B have perfect integration within C.

The diagrams in Figure 16, are very general and apply on all scales—including representing the computer bit. Figure 16(a) essentially sums up our 3D level of thinking and science; a world made up of parts (A and B) stuck together by forces. These also correspond to the bits in our linear computer. In Figure 16(b) the triad principle C is the key to the greater truth. As applied to the mind computer, the triad is the computer 'bit' and these triad units can be built up indefinitely.

Thus the Absolute clearly cannot be divided; its very definition prohibits that. It can, however, counteract itself. As the division continues down the fractal levels, the A's relationship to B's becomes more separate and A and B become more objective relative to one another. One side can be an environment; obviously structured very differently* (but remember even an atom has a speck of the Absolute), which has inherent within it consciousness and the aliveness characteristic. Both the internal (subjective) and the external (objective) are created from the top. See Figure 15. The objective side is formatted into fixed (no free will) universal components, such as an atom. Thus all energy is originally an aspect of consciousness of the Absolute (the basic aliveness characteristic). The objective environmental part of the division must be unconscious to the observers (such as us). [*This is why quantum physics revealed that the observer (even in the scientific experiment) was part of the set-up or environment.]

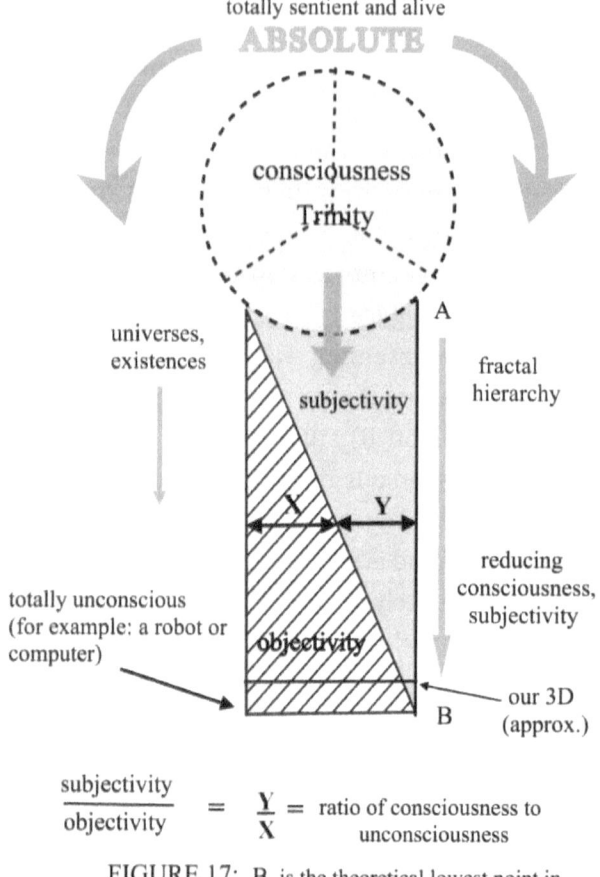

FIGURE 17: B is the theoretical lowest point in practice (but note that mere atoms have a degree of consciousness, that is, the Absolute).

Thus the two opposite intentions in the counteraction simulates an energy effect for stability and permanence and the first principles for the illusions of 'solid' structures. The two new parts must be in a different state from their primary condition of the Absolute (different degrees of order/dimensions), and this different state keeps them all apparently separated, meaning they are separated with respect to the context of the divided state/level, but that this separation is an illusion of objectivity, since the parts are one whole at the primary level or source. This would also introduce the elements of space and therefore time, plus basic energy, and these two divided opposite energy flows can be considered to be the beginnings of created elementary mass.

Now on the basis of what we know about wave forms we see that the appearance of *existence* would apparently arise from the balance between wave forms; that is, a delicate balance between *stability* and *instability*. Instability is not a negative factor but allows the change in the stability (which otherwise would result in stagnation). This is the essence of the paradox between expansion and consolidation, which we have covered previously, particularly in volume 1.[2]

For time (a time line) to exist and a sense of duration, there must be repetition of this procedure. We then have frequency. Something regularly turning on and off will give a sense of objective time.

As we move into manifestation from the primary Absolute condition, the triad principle gives a certain freedom from the Infinite by becoming finite. The reference to Spirit would relate to the first unified field of energies.

Let us recap and delineate some fundamentals of the Creation process:

1. The original basic state is the Absolute: neutral, unconditional, eternal.
2. There is only consciousness, manifesting in different forms. In its primordial state it is the Absolute. Inherent within it are the non-quantifiables: the aliveness characteristic, basic sentience, and the experiential.
3. The Absolute observes itself by counteraction and focussing on one side of the division (A and B, Figure 16), creating not-knowing, objectivity and unconsciousness of the other side, say, B, where both A and B appear on a lower (dimensional) level.

4. This Absolute can only be totally conscious and a single whole (see Figure 15), which is totally subjective.
5. The Infinite Absolute is converted to finite by the application of the triad principle: first the Trinity, then further divisions for lower dimensional levels, universes and existences.
6. The Absolute, basic consciousness is in everything and all of it is everywhere at once. Through the Absolute this means everything is everywhere at once; a property of the Absolute but can only have meaning *when* the Absolute is in contact with the space and time features of its creations/manifestations. (Ocean analogy: no matter what wave patterns there are in the ocean, the water is always present.)
7. Life is a delicate balance between stability and instability.

An understanding of the learning pattern would convince scientists that the linearity of a particle and objective world have no real basis; they all fail the tests of truth of physics. All creation must come in nonlinearly and they must always be interconnection between apparent separate elements of existence. There is only control via nonlinearity. The physics of the learning pattern also reflects principles on a cosmic scale. The hologram is an interface mechanism between the whole and the part, bringing about the formation of holographic fractal dimensions for exploration of consciousness, giving free will to lesser parts. These more limited parts may then go astray and the negative and dark side of existence arises. The raw unconditional Absolute will accommodate this. Just as in the ocean analogy, the water is harmless but something could put a turbulent pattern into it (but the power exists always and only in the water medium).

Finally, can we bring any science into the religious belief of how creation occurred; that there was the word of God, and then there was Light? Yes, if we had a science of the mind, this could have been understood. Briefly, 'the word' refers to the sound.[3] Science hasn't yet related sound to magnetic and scalar phenomena. How about 'then there was Light'? Light is electromagnetic and here it correlates with electricity. Thus we now have the two principal ingredients for all Creation: electricity and magnetism (though we might note that at this highest level, magnetism came first). Even more surprising, a science of the mind will find out that the male mind processes mainly electricity (with a small percentage of magnetism) and the female mind processes mainly magnetism (with a small percentage of electricity).

The scientific 'Trinity' would be the initial unified field (compare Holy Spirit), and the 'Father' and 'Son', the magnetic and electric entities (there are strong indications that 'Son' was originally 'Sun' * (hence 'Light' and electricity). Note that this is not weird when one realises that all manifestation has the Absolute within it (the aliveness characteristic). For example, in the vast machinery of multiverse creation there will be transformer beings. [* What could be the reason? Clearly, introducing 'Son' (masculine) was all part of the campaign to emphasise the masculinity (note 'Father', the magnetic and feminine) and make women submissive.]

Notes
1. Book: *The Emerging New Science* vol. 1.
2. Ibid.
3. All manifestations of energy and matter are basically wave patterns, as per quantum physics. All wave/frequency patterns contain a sound and light component. The sound part relates to the creation of form. Thus specific word sounds could create the frequency patterns, and subsequently materialisation.

14.

ARTIFICIAL AND UNNATURAL REALITIES

Science does not distinguish between natural reality and mutated reality. It detects the difference but fails to see the significance.

Our civilisation is so thoroughly programmed with false data that our collective is creating false realities. This is due to information fed into the minds of individuals; each has potentially an enormously powerful unconscious, and collectively this power can act like a god, regarding creating reality.

If one wanted to enslave a civilisation, a potent method would be to separate the subjective from the objective. However, you might know that the subjective and objective cannot be divided. They are like two sides of the same coin.

This can be overcome by causing this separation to be achieved artificially, resulting in an artificial reality. We then form two artificially biased subjects of knowledge: science and religion, each is consequently a partial quantum reduction from a higher context. Science endeavours to become totally objective—completely paradoxical since objectivity, as we have seen, is an illusion. If everything truly became objective, everything would disappear. On the other hand, science forms a dichotomy with religion, of which the latter becomes almost entirely subjective and experiential, based on faith and beliefs. Each is set up in a 3D polarity relationship with the other, and if we try to unite them, they will destroy one another. The hidden context must be found to remove the conflict, which brings them into one subject, and then the higher context is identified, which is a single subject: say, a 'spiritual science'.

Thus the individual compiles knowledge based on objectivity and separately, subjectivity, and the relationship between them is never understood—this is the situation that exists today. Universities should have a course teaching an understanding of the relationship between subjectivity and objectivity. If this had been the case there would never have been the forty-year debate between Einstein and Bohr, referred to today as the greatest scientific debate ever (this was covered in volume 1).[1] This is thus an example of an artificial reality. In general, we have the natural reality plus the non-organic, spiritually misaligned, unnatural existences, which we can divide into two further categories.

1) Normal, organic, natural reality.
2) Reality based on malfunctions of energy and mutations.
3) Artificial realities.
    (a) Internally applied mind programs.
    (b) Externally applied holographic inserts.

Numbers (2) and (3) above will inevitably overlap to some degree.

In category (1) above we have a planet and environment with naturally functioning energy interrelationships. It is the ideal scene in which all processes, whether behavioural or technological, are in alignment with the holographic and harmonic (Divine) cosmic blue prints. However, let's move on to (2) and (3) and come back to (1) since this section is mainly about the artificial systems.

In category (2) we have emphasis on malfunctions and mutations. This refers, in particular, to the human DNA (also the DNA of lesser species), in which over 95 percent is missing. Science does not recognise this and classifies the missing DNA as 'junk'. Also within the external universe we have malfunctions of our planet, and particularly the whole solar system (science recognises this)— wrong orbits and spins of planets, etc. Note that Earth is tilted 23.5 degrees and although it creates the seasons, this is not natural. As we shall see, all bodies are formed by the dual-vortex system and that these are supposed to be in alignment up to the highest orders of the cosmos. Earth's tilt puts it out of alignment with the solar system, specifically the Sun. These are not accidental or random faults—the universe system won't fail in itself—they are caused by interfering galactic races, very advanced technologically but deficient spiritually, with little regard for other life forms. Scientists have recognised the black hole at the centre of our galaxy but not realised that this is not normal and in fact it means that it is slowly compacting instead of expanding; hence as a result it is a degenerating galaxy.

In (3) above we have artificial realities formed internally (a) and externally (b). By 'internally' we mean the human mind is programmed to believe certain artificial conditions are a reality, resulting in the collective consciousness bringing about the manifestation of this reality. Even with a belief that, say, economical difficulties are to be expected and are quite normal, this normality will be automatically created by the civilisation. Another belief is that life comes from matter and then teaching theories and conditions that appear to prove that. A common one is that time is age, and that time naturally ages anything.

The really clever artificial reality is the one mentioned above, since it can be utilised to enslave a civilisation, which is to separate knowledge into science and religion. This dualism does not exist in the cosmos and, as we have seen, it is anathema in the 'tests of truth' of physics. This abnormal dualism is achieved by separating subjectivity from objectivity. The reader may still, however, protest that they can't be divided. That is correct, in a natural and aligned reality, but artificially it can be achieved when the relationship between subjectivity and objectivity is properly understood.

Thus the control mechanism here is to artificially separate subjective knowledge from the objective. Contributions are from science and politics, and in particular by education, rewarding and developing the left brain characteristics (intellectual, logical, emphasis on the parts and objectivity) and subduing the right brain abilities, such as intuition, imagination and subjectivity. In fact this manipulation goes much deeper, influencing evolution itself. Over 95 percent of our DNA is missing (is not junk DNA) giving rise to a mutated carbon body with two brain lobes instead of one. These two formatting systems (like filters) cause separation of the basic consciousness/mind into the characteristics mentioned above. Now these two attributes (intellect and intuition) are manipulated by the educational system, that is, reinforces left brain and suppresses right brain. It just so happens that the left brain, no matter how brilliant it is (intellectually) it can be programmed much more easily than the right-brain intuition (which the latter in itself can't be programmed since it acts only in present time, a single whole quantum state in the now). Science doesn't recognise that true CAUSE is in the NOW, and feedback from memory, learning, all imprints, are EFFECTS in the NOW. A robot does not do anything in the NOW (its information already exists, only the effect is in the *now*).

Both these subjects, science and religion, are partial quantum reductions from a single higher context, and each subject in isolation can't ever be complete. If science uses the subjective side in making evaluations it will simply 'collapse the wave function' (quantum reduce) from a higher (more coherent) truth to its own degree of order (that of the experimental set-up); and if religion utilises the rules of objectivity, lack of belief starts to emerge. The ladder analogy given earlier in the book reveals this by showing that the bottom level of existence is all that science deals with, whereas religion aspires to the higher rungs and, in particular, the source at the top.

In category 3(a) we have artificial realities caused by programming the group mind or the collective (which could develop the power of a God—recall the process of quantum regeneration), whereas in 3(b) the reality (for example, objects and events) are externally created by inserting material features into our life or time line called holographic inserts. The most famous example, as described in the author's other books is the event of the resurrection of 'Jesus.'[2] It was in fact a holograph of another preacher personality by the same name (the one who actually married Mary Magdalene and whom the Romans mistakenly took for the real Jesus). But there is much more to this complex series of events; in fact, incidents in the lives of three personages were merged (one was Jesus).[3]

By separating science and religion into objective/subjective categories they become in polarity conflict at the 3D level. In trying to unite them (in 3D) they will destroy one another, since their relative and contextual aspects are obviously incompatible. It is necessary to find the hidden context that separates them and then the higher context that unites them (this was described in volume 1).[4] One may see the extent of the manipulation that is possible and is occurring within the human race.

Returning to the natural reality, we have drifted so far away from this that all sense of purpose is lost, in particular, evaluation of importances in life. Restoring the natural reality is, of course, not separate from an understanding of the human's true history, continuous life, etc.; also what constitutes the human being, which we dealt with in earlier sections.

Leading scientists for more than 50 years have recognised that the universe is holographic. This, combined with the fractal design of dimensions of different degrees of order for the exploration of consciousness, directly relates to the presence of harmonic blueprints. If we simply take the hologram we can see that all degrees of order within it are cooperating, in harmony, resonance—the whole to the parts relate; no one part is obstructing another; similarly for Cosmic Creation. Life comes into the system and is given free will (there will be different degrees and regulation of this). Eventually life goes astray and deviates from a behaviour based on harmony with what could be called the Divine Blueprint. All energies interconnect, consequently any harm done to another region or life form causes out-of-phase energies (sine waves out of mathematical alignment) impairing the harmonic and 'benevolent' energy flows of the universe, creating blockages. The Cosmos handles this with the rule: To every action there is a consequence.[5]

Thus built into the Cosmos are override mechanisms, immune ('antivirus') systems, and of course karma.[6] All is expressed in wave patterns (quantum physics) and anything negative will radiate wave/frequency patterns, which are the resultant of sine waves out of phase, out of mathematical alignment (creating spiky resultant waves). These are radiated out into the universe (as per quantum physics). The karmic action logically appears to function by returning to the individuals the negative wave form they put out in an attempt to cancel it by phase conjugation (if one kills someone, eventually the action comes back to one).

There is, however, no magical cure, the *external* action of the returning wave form will not work in a purely mechanistic manner. The state of mind of the individual will govern whether the negative recording erases. This means that consciousness must participate in this karmic process by recognising what is happening at least subconsciously (a little like 'turning the other cheek'), and avoid resetting the negative state of mind; there must be no resistance, revenge, vindictiveness, blame, etc. This is almost impossible in most cases for the human on this planet, and as a result, karma, in general, is not working (by this cosmic method). However, clearly the individual is ideally supposed to handle the negative wave pattern by self-improvement, or specific healing remedies, etc., *before* the 'cosmic consequences of the action' occur.

Thus the basic energies of the cosmos are set up for benevolence, integrity, ethics, responsibility, etc. They are part of the natural structure of the design—as mentioned above, regarding the hologram. The system promotes cooperation, behaviour and actions for the greatest good. This is what 'natural' means: living an existence that is in alignment with the Divine purpose (doing God's will) of exploring consciousness and ascend by expanding (more 'wholly') consciousness up through the dimensional fractal system of increasing degrees of order, back to the Source (ultimately after maybe millions or billions of years).

Notes
1. Book: *The Emerging New Science* vol. 1.
2. Book: *Engaging the Extraterrestrials.*
3. Book: *Voyagers* vol. 2 by A. Deane.
4. Op. cit. *The Emerging New Science.*
5. Workshops on the Guardian material by A. Deane.
6. Article: *The Mechanics of Karma.* www.nhbeyondduality.org.uk.

15.

THE HEALING PROCESS

No healing occurs in 3D but there can be assistance from belief, symptomatic treatments or catalyctic actions. Only restoration and alignment with the natural harmonic blueprints and templates heal.

In this section the difficulty the reader may have is in understanding the difference between external and internal processes involved in the healing or treatment techniques. The population has been educated and programmed to believe that existence is only within the external 3D. This is the domain of the scientific measurement, requiring an experimental set-up with a stringent (100%) objectivity in acquiring knowledge on how the universe works, which we have shown is only a limited view of the real universe.

Mainstream medical knowledge and science, generally, teaches us that the human body is a three-dimensional structure, which when subjected to a lifetime of environmental interactions and simply time itself, wears out, somewhat as does an old car. This is commensurate with the medical profession treating the (external) symptoms of illness and not the basic (internal) causes. For example, the symptom of the presence of a tumour is not the cause of the problem; it is merely telling one that there is a more hidden fundamental dysfunction that needs handling—usually mind-related.

The patient, whether under psychiatric treatment or medical, learns only intellectual external information about their ailment; meaning they are not educated to believe that they, in mind or body, basically are the healing agent. Anything external, if it is medical, may handle the symptoms but the problem may return or reappear in a different form. However, other personal factors may be present, such as a positive state of mind, a belief in obtaining results, or even an imminent karmic release which, say, the mind has almost resolved, and thus the treatment, even if external, may act as a catalyst for the body to resolve the dysfunction successfully. It is clearly a complex mind-body matter.

For a proper healing, not dependent on blind faith or karmic completion, consciousness must participate in the process. In the case of mental healing treatments, any intellectual description the psychologist gives to the patient is only external information and will, if anything, cause the patient's consciousness to look at their problem externally, distracting the direct inner interaction for a proper resolution. However, the patient, in general, has no idea about the internal process.

In addition, it is not surprising that the body is so vulnerable and prone to dysfunctions, since there are fundamental flaws in this carbon body. Let us take the above car analogy in the opening statement of this section, comparing the problems the old car has to the human's state of health. Imagine taking it to the garage and asking them to fix everything. We return to pick it up and are told that everything was fixed that could be fixed, but that the car had inherent faults in its design and we would have to tolerate these basic flaws (or buy another car). We are talking about mutations resulting in an unnatural carbon body. This topic of how the mutations came about has been covered in previous writings.[1]

Illness, science will find, is far more complex than is depicted by current medical knowledge. This is not surprising since science does not recognise the true role of consciousness, as indicated throughout this book—this is also why the effects of belief and faith are a mystery to science. Without conscious participation of the patient or any apparent belief on his or her part, cures won't generally take place, except in the case of established standing medical practices where people have been thoroughly informed and indoctrinated since childhood that only the medical system can cure. This becomes a programme in itself and becomes embedded collectively. Thus there is in fact a kind of collective automatic belief that medicine works—an inherent agreement.

This is similar in principle to the mechanism of karma, which won't work without the participation of consciousness; or to the suppression of symptoms with drugs; basic consciousness is not participating in the healing. Generally, medical treatments handle only the symptoms (not achieve a basic cure), and exceptions are when the belief systems of the patient were sufficient to effectuate a cure via the symptom, which is connected to the hidden source of the problem. In fact, epigenetics, which recognises that the DNA is influenced by environmental factors, including thought, very much supports the healing process originating from the state of mind of the patient.

Responsibility for self is the key to begin the healing process and will tend to function more readily when the individual is involved consciously in the healing process. However, it is not necessary that one should have the knowledge of a medical doctor for the healing process to take place, since one's treatment can utilise systems of healing, including medical practices, all of which can act as a catalyst in the self-healing of the individual. In fact, the source of all health problems will have some kind of origin, beginning with consciousness (expressing as mind).

The illness will always be accompanied by a degree of contraction of consciousness in some region, such as from a negative wave/frequency pattern out of alignment with the underlying harmonic wave patterns. Thus healing involves a restoration of alignment and increase in consciousness, which is also the essence of a progressive evolution (that is, the same mechanism: expanding and increasing alignment of consciousness with the Source and Cosmos).

People are programmed to believe that only certain practitioners can heal them, or only certain established medical procedures will work. Intention and desire to heal is most important. We hear of 'fighting cancer'. This is not a good attitude since it reduces the communication within the individual's own healing process—it becomes more external: one side of the communication line blocks off the other. Anything of a conflicting nature is not conducive to healing. When true healing occurs there will be a change in the person's mental state. It usually involves bringing experiences, feelings, to the surface. However, these healing reactions may sometimes result in a temporary worsening of the condition before benefits ensue.

Why is the placebo so effective (for example, a 'sugar pill')? Such a deception is not a logical process for a cure but if a person believes in it, it may bring about a cure. The belief enables the person to participate in the process, and since belief is an inherent component of the creation process, it can work. A more scientific and rational system with an explanation for healing may apparently heal, but it may only handle the symptoms and not the cause of the problem (the external system cannot heal of itself permanently); the problem may return or take another form. It is very complex; there are many interacting variables that can influence the outcome. Nevertheless if the person has a strong belief, it could result in a permanent cure. What are the mechanics of this?

The more one recognises a logical and scientific system for the healing, the stronger the belief factor (for those persons requiring a logical explanation). Ideally, healing would occur spontaneously, that is, the primary creative cause operating within the natural order of the human body, blueprints and templates would heal the body and mind directly, but our attention/awareness tends to move to the external component of the feedback loop since we are educated not to believe such magic. Recall the automatic awareness and the cybernetic feedback loop in Section 10. If the feedback is over-controlling its source, the primary flow, it will block the healing process coming from the internal primary flow, the spontaneous healing. Nevertheless the belief or faith factor is naturally aligned with the control side, and it would not be impossible to program a positive state of mind (into the primary flow) to assist or achieve healing (this doesn't constitute an intellectual understanding of the program that we have shown will hinder any cure).

This feedback effect applies generally. The individual thinks the environment, the external view of the observed, is truly objective—that we (1) have no control and are the effect of the observed and have lost sight of who we are. But ironically this belief in the external (such as a placebo) can trick the subconscious so that it heals the ailment spontaneously but it (the subconscious) 'thinks' the external and substitute path of healing is doing it. The subconscious, being tricked in this way, tends to keep in de-activation all the counter intentions, negative thoughts, brainwashing, thinking it can't heal, enabling the external method to follow a clear path to apparently heal.

Therefore the external system, which doesn't heal anything on its own, will require a degree of faith (to bring in the natural healing). The greater the faith (for a particular method of treatment), the more likely it is to heal. Or, contrarily, the more logical and efficient is this external treatment, the less blind faith is needed to obtain results. And in this case, by logical, we mean a process of treatment which aligns most closely with the spontaneous natural healing of the original divine blue prints.

When the system is more correct and in phase with the natural order, it is easier for consciousness to follow a path of healing automatically, and belief is taken for granted. Thus with a less 'correct' treatment, the greater will be the faith required. However, few people in our day of mechanistic educational knowledge would demonstrate this faith and belief-dependency. Nevertheless faith in the external treatment as being the cause, may enable the subconscious to heal.

A ritual is a procedure in which power is assigned to certain actions and symbols, enabling the creative energy to operate on belief—devoid of material connective paths between cause and effect. There are exceptions though where certain movement patterns, such as a dance, resonates with natural energy flows or templates. We are trying to convey details here of an overall model as a guide to understanding these rather complex concepts. Consciousness, in general, is continuously splitting into the 'I' (self or observer) and not-'I' (environment/observed) modes. The not-'I' structure of matter, energy, space and time, by feedback, acts like a filter or mould on the 'I', that is, consciousness takes on the not-'I' structure compulsively—the not-'I' focuses the 'I' into its (the not-'I') configurations (remember the environment is an extension of our unconsciousness). Basic consciousness still is responsible for the work but it is manipulated and moulded by its own structures unaware that it is still the cause. The structural part both in the mind and environment contains the external path. We are stating that there are different degrees of this path according to different activities. The recognised miracle of faith is observed as containing no intermediary or external path—it is at the top of the scale. The faith may be in some symbol, even if it is the image of, say, Jesus Christ. Whatever the individual assigns the power to, will enable it to cause the 'assigned' to appear to work, but in fact the healing operates unconsciously. As we move up the scale (into higher dimensions of existence), objectivity and enforcement of structure on consciousness diminishes and there is less power from the universe to affect us.

A further possibility for healing is that the medical system has been established over a long period so that knowledge of success is engrained into the subconscious of everyone. There is then a collective belief, without any necessary consideration of individual faith, and cures may result.

Even the universe of space, time, and matter itself is a representation in an objective format. It is a mode of expression of consciousness. The lower part of consciousness' spectrum separates the most from itself, forming the ego view and denies responsibility for the universe's existence, that is, for consciousness' participation in the universe's creation from the quantum state or the Infinite's energies.

Thus just as we have internal and external in the healing process, so similarly universal energies are not separate from us as governed by the 'I'/not-'I' relationship. Radionics and early psychotronics have a position on this mind/matter scale; they are above the position of present material science and machinery on the scale. Their hardware of electronic components is similar, but because their design is not fully scientifically defined or complete, the mind must play a part and, in effect, the equipment acts partly as a substitute path for the mind, compared with normal technology in which energies are more objective (but note that psychotronics has been developed effectively by the shadow government). The element of certainty, agreement and faith will be required to operate them. However, as stated, science has secretly progressed and such psychotronic equipment will have become more objectified and established (and a better duplication of the natural energies to create the result) and it will work even without the individual's conscious faith.

Thus on the lowest level we have our common machinery based on classical physics, such as the automobile engine. This is at the bottom of the scale of increasing objectivity; every detail of the workings of the engine has an explanation of a 3D nature and is established and agreed upon. It will still work even without individual belief—though the individual automatically believes in it. It is, nevertheless, a structure of consciousness extremely objectified (it is a product of the denial part of the feedback loop). This means the engine is disowned as part of consciousness, and the machinery still functions. The mind does not have to play a part as a 'free agent'—in a creative manner—it simply 'obeys' existing programmed structures; it is being/duplicating these structures compulsively (by feedback). This degree of objectivity—matter, machinery, etc.—will enforce its objectivity and 'faith' onto the observer. One is 'forced' to go along with the physical universe, the existence of objects, physical events, etc. They (templates of manifestation) mould their activity and appearance within the medium of consciousness.

This is all not so different from the relationship between the computer scientist and the computer on the desk. The scientist created the computer in the first place but separates from it and allows it to function to some degree on its own and provide (feedback) useful information for the operator.

An example of possibly the best external treatment (meaning the treatment procedure has a complete explanation scientifically but is also most aligned with consciousness) is the one developed by Priore.[2] The problem area is scanned and frequency patterns recorded (containing dysfunctional patterns). These frequencies are then radiated back on the problem area but in phase-conjugation mode ('upside down') to cancel the negative wave patterns. Consciousness will more easily participate in wave-pattern systems since this is closer to the core of creation. However, to increase this efficiency of creating compatible systems, which are external, relative to the natural healing, the Priore technology also involved scanning the whole body with the same pattern—thus allowing for the fact that basically everything natural is holographic (the negative pattern will also be imprinted within the whole). This particular healing treatment will therefore require minimal 'faith' to be effective, in contrast to, say, a placebo at the opposite end of this scale.

Nevertheless even with this system it does not heal; it cancels out the negative pattern by removing temporarily the negative influence, thus enabling consciousness to participate in the healing process relatively unhindered.

Now, the healing process is complicated by the fact that the mind not only records every detail of events and experiences of the present life, including emotions, but also from past lives. It is astonishing that science has not recognised that it is elementary physics that the mind continues on after the death of the physical body with all abilities and memories (in actual fact this is controlled and suppressed knowledge). The mind is a quantum field of vibrations: the vibrations don't require feeding and don't wear out. Thus past-life traumas can be reactivated in the present life giving rise to all kinds of somatics, ailments, and mental illnesses. Note that all these reactivations from the past are part of the feedback of the cybernetic loop. Psychotherapies can handle this by procedures in which the individual is regressed into the past, then running one's attention through the incidents repeatedly until the negative emotions discharge. A resolution will be accompanied by alignment of wave patterns.

Thus a past traumatic incident, resulting in pain and injury causes consciousness to withdraw from the recording of the event and to resist the experience—rejects the feedback (see Figure 10). This creates a counteraction of wave forms with misalignments (of 'sine' waves) and stuck flows in space *and* time. This is now dormant but ready to feed back the negative recordings when activated by similar events in the present. When such incidents are brought up to consciousness, the latter forms a separate pole, and the pain stored as electrical charge, as the other pole, can be discharged permanently.

Note, however, that if the traumatic incident is too severe to confront when recalling it, the discharge will not take place. Consciousness can't confront it. The therapist must have the patient recall more recent similar experiences of a less severe nature and clear those first;* this will lighten the load on the original incident and it will subsequently release, and the energies re-align.[3]

* [Failure to do this, which was common in psychoanalysis of the past, has resulted in suicides—a textbook of the author when doing a doctorate in psychology stated that 15% of patients undergoing psychoanalysis in the previous year committed suicide.]

Thus no external systems heal permanently. They can, however, act as a catalyst and aid one to heal oneself. Only our consciousness and body heals us, not something external. It is so easy to put one's attention out on this and subconsciously one is waiting for external help to take effect—it won't. When the external help seems to work it won't generally be permanent. Thus the conscious mind must participate and take *the* responsible role in healing (but as described above there can be an automatic or collective faith accompanying a treatment).

It's like karma. There are three conditions arising from this. If one has incurred karma, then 1) self-improvement in the appropriate areas of behaviour can heal the negative wave patterns (the karmic pattern) and avoid the universe's karmic mechanism designed to remedy the fault, which is by returning the flow; 2) if this self-improvement does not occur, the universe returns the karmic behaviour back to the person—this return appears to be the same negative wave pattern 'upside down' (technically: phase-conjugation, or 180 degrees out of phase, or time-reversal). If one recognises this is occurring (that is, learns something from the negative effects) and consciousness participates positively in the experience, the karma is cancelled and the negative experience will not repeat; 3) if there has been no self-improvement, essentially in the karmic area, and in addition, one doesn't participate in this external potential healing process, it will go on repeating indefinitely. This is the major condition today on Earth, and thus karma mostly isn't working. People are not recognising their role of responsibility—we are taught that power is external and not internal.

In effect, the paths from the healing energies must come from within one and then if necessary combine with the external for assistance (a proper direction of flow). This is in essence a cybernetic process described in Section 10.

Notes

1. Book: *Engaging the Extraterrestrials: Forbidden History of ET Events, Programmes and Agendas.*

2. Priore article. *Suppression of Knowledge, Discoveries, and the Free Energy Problem.* www.nhbeyondduality.org.uk.

3. System of treatment using an E-meter, taken from Scientology philosophy.

PART THREE

SCIENCE IN THE ARTS AND SPORTS

16.

SCIENCE AND THE ARTS

The more one understands the common denominator of different subjects, the higher the context and level of knowledge attained.

[Reference: abridged from articles: *Understanding Art, Part I, II and III*. www.nhbeyondduality.org.uk. N. Huntley.]

We shall mainly focus on the visual arts but will include music; in fact, music might be considered as art in time. We are particularly interested in art and music from the scientific point of view, in other words, any basic relationship between art and science.

Let's begin by pointing out that today the arts are underestimated in value and intelligence on this planet. We have an authoritarian hierarchy set up in which the more mechanical intelligence, logic, intellect, appear sufficient for the development of a civilisation; in fact are the criteria for dictating and running all facets of practical existence and society. This is a deviation from the ideal. However, an analysis of art and its experiential aspects is immensely complex and this is due to the fact that the human is immensely complex.

As stated previously, physics is the basic science and its application is relevant to all activities. The logical and intellectual are a manifestation of the quantitative and linear aspects of existence: separation achieved by parts out of phase with one another (different frequency). Quality involves in-phase, resonant parts that then create true unity (holistic, holographic states), not parts stuck together by forces, which is all that science recognises today. Quality or qualitative could be defined as meaning the degree to which all inherent parts are in phase, that is, resonant.

The arts, whether visual art, music, poetry or dancing, communicate quality (even dancing, or movement, in general, creates energies—patterns of frequencies).

We use the word 'quality' somewhat loosely in everyday life, for example, quality of materials, or tools, or life itself. Even so, the definition could still have some degree of application here: less impediments or impurities, parts more harmonious, etc. These are all factors conducive to greater integrity and longer duration.

Let us recap on some of these terms. There are two fundamental variables to be acknowledged in our existence, which are categorised by 'quantifiable' and 'non-quantifiable'. Science only recognises what is measurable, that is, the quantitative elements. Now 'quantifiable' very much relates to quantity or quantitative, and non-quantitative with quality or qualitative.

As far as science is concerned there are three classes here to gather data from: 1) the quantitative (which can always be potentially measured) or 2) the qualitative that arises from the special cases of the quantitative (in which the parts are in harmonic relationship giving rise to quality),* and 3) the experiential aspects of the qualitative, such as in art or music. [* Recall quantum regeneration of parts into a coherent state (qualitative).]

The qualitative descriptions include unity, wholeness, and resonance of frequencies. These properties are commensurate with higher truths, which relate to greater integration, coherence, etc. Science should be handling this; it is well within its field of capability. However, science cannot cope with the experiential and will never measure, detect or analyse the non-quantifiables of life (the basic Absolute: sentience, pure consciousness, and experience). The point we wish to make here is that science should be explaining and evaluating not only phenomena in category (1) above, but also (2). However, it will never evaluate the experiential since this is an Absolute property.

Recall again the analogy given previously of the geometric world of the circle and the straight lines (such as imagining structures built of matches). The straight lines are the quantitative and the circles represent the qualitative. The match-stick structure can never fully duplicate the circle or curved lines (the experiential), but can represent approximately a circle and become more qualitative. A practical example could be by examining the frequency patterns of works of art and recognise differences of merit, such as good, bad, by detecting the expressions of frequency patterns created in the appreciative viewers mind (measured with an imaginary scalar electromagnetic oscilloscope).[1]

We shall also see that the characteristics of these energies run parallel to the path of higher levels of intelligence, frequencies and fractals of consciousness. Qualitative energies link up to the higher fractals and aspects of consciousnes, including intuition, which has the ability to access higher aspects of consciousness. All these factors, quality, context, fractals and expansion have in common the same energy structures and utilise the same physics.

The laser mechanism is an excellent scientific example of the conversion (quantum regeneration) of non-harmonic parts into a coherent state. See Appendix B. We begin with quantity; for example, a number of rays of white light. Quantity (as opposed to quality) means the parts are separate, unrelated; the rays of light are higgledy-piggledy, are out of phase, that is, random. If there are, say, ten rays (for simplicity), in the case of ordinary (white) light, the rays will be out of phase, that is, random—not aiding one another. The total energy from these ten rays would be about the same as three of the ray's energies added together. Now when we put the rays of light into coherence so that their oscillations are in step in space and time—this gives the laser beam. The rays of light, quantum regenerate a higher-frequency oscillation and the total energy is more than the sum of the rays; it is ten squared (100) plus. This huge amplification applies for all groups in resonance.

Let us analyse the quantitative and qualitative aspects of a good work of visual art. The bits of paint on the canvass are quantitative, physically, in that they are fragments stuck together by forces. However, the artist has placed the elements of paint in a special order such that the bits of paint can be correlated. Thus, the artist has created a qualitative effect from a quantitative state or medium. The appreciative viewer now looks at the physical surface of the painting; eyes may go slightly out of focus, the intellect and conscious mind step aside and the 'unconscious' now correlates all the bits to quantum regenerate one whole state. Because the bits of paint (the 'quantitative') have been subtly ordered and interrelated, then resonance can be achieved, resulting in a single whole energy that has (hopefully) complete unity and is a quantum state—a single frequency pattern. There is really no 'art' on the canvass only perception elements which interrelate, that is, quantum regenerate when viewed by a higher-order observer.

This condition of art appreciation, of course, occurs in the mind but is not the actual mind recordings. We are talking about higher aesthetic effects of the input consciousness (recall robot and input analogy). The result of the appreciation is experiential and can only be the property of the Absolute even if 'caused' by the imprint/-template/mould. Recall earlier what this input consciousness consists of. Thus a qualitative state in the mind of the observer is now superimposed on the quantitative state of the paint on the canvass—and appreciation has occurred. Of course we can't say much about the experiential aspects of this in terms of physics, except that it will be at least pleasing, will have a sense of unity, harmony and completion—nothing need be added or taken away (depending on its merit).

Note here, as already indicated, that the integration and higher spectra aspects of good art and music involve a physics that runs parallel to true evolution itself, which is mainly expansion of wholeness and developing higher frequencies. This would become evident if we could display on an oscilloscope both the wave forms for art and evolution.[2]

What is the mechanism of appreciation in music? Imagine listening to a good piece of music; we only hear one sound at a time (even if it is a complex chord). There is clearly no music in an isolated sound (science is with us on this). There is no music in a sound 'now'. The mind holds on to immediate past sounds, anticipates future sounds and combines them as one whole, past, present and future over a short interval of time. Science has no argument with this. However, the process of perception of these sounds—past, present and future—creates, science says, the *illusion*

New Science

of the tune/melody; the essence of the music. This is the emergent software. But it is not an illusion. It is a real energy/quantum state: a frequency pattern; a quantum-regeneration. Science can't detect these high frequencies (we have covered this thoroughly in earlier chapters).

This energy state, the original idea, is what the composer created and then had to quantum-reduce it to parts—the notes on the music score—as efficiently as possible. The listener then re-creates this aesthetic energy state by listening to the parts, then connecting together (not intellectually as science thinks) the past, present, and the future anticipated sounds, over a sufficient interval of time. This unconsciously then resonates with the whole essence of the music, assuming the music has sufficient integration/wholeness (a simple good tune has wholeness: a single quantum state of energy, a frequency pattern).

How much is the music dependent on the listener's role? As stated, there is no music in the sound 'now'. What is the mechanism that enables one to correlate the parts in time? This is the purpose that rhythm plays. In effect, it acts as a space-time framework to hold the musical meaning and to assist the listener in grasping the musical context. How does it do this?

The role of musical rhythm qualitatively is not generally understood. We understand rhythm as a strong, regularly repeated beat or pattern of sound, or movement. Clearly it is not in itself music but a temporal system for carrying or conveying the primary essence of music, which at this point we can think of it as musical aesthetics or the intrinsic aesthetics of musical essence.

Anyone may have observed that a sound, repeated regularly in time, creates anticipation of that sound. In fact everyone is familiar with this in noisy or unpleasant sounds. The particular anticipation increases the irritation. The regularly repeated sound also tends to create a hypnotic effect. The repeated sound tends to have a certain hold over the mind. What exactly is this?

We mentioned above that in musical appreciation, we remember the past sounds, combine these with the present ones being heard now, and anticipate future sounds based on the past and present. But in science it just means 'association'—two things 'stuck' together. This type of system is a product of intellectual analysis and is fictitious—compare Figures 16(a) and (b), Section 13.

In the same way that science fails to recognise all true wholes, such as in art, or even the whole underlying quantum state of an atom, planet, star, etc., it doesn't recognise that in the process of musical understanding, the past, present and selected future are merged into or by a whole (4D) quantum state. This is a small interval of time, but extending into a few notes, which is spanned simultaneously. This interval (a quantum 4D state) will have no time in it; it is like a larger unit for time—compared with objective time (which is like a clock ticking out small units of time). In quantum physics this unity or quantum state in time is called quantum action.

It is necessary to realise that while appreciation of a musical sequence is taking place, that is, attention spread over the interval spanned, the listener may instantly change this focus of consciousness to, say, an intellectual, analytical focus. If one takes an interest in, for instance, the sound of a particular instrument, the quantum state of appreciation (which is a wave pattern/function) will 'collapse' (the so-called collapse of the wave function) and the mind will then be giving attention to the parts, such as the single instrument, or how well it is being played. The state of appreciation may be flickering on and off as one pays attention to other things. There is a flickering from the whole to the part (the whole collapses to the part—quantum reduction—and is then quantum regenerated back to the whole). This is fine and normal.

We may note then that there will be variations in ability to span time in this way and therefore variations in degree of appreciation of music. A simple pop tune may be easier to appreciate than a more complex classical musical composition. We can see now that the hypnotic effect of the anticipated beat of the rhythm assists a person in spanning sufficient time to provide the basis for recognition of the musical aesthetics—the tune, melody. Sufficient attention span is required for appreciation; enabling understanding to take place for a particular piece.

We may observe that animals have low attention span (stimulus-response learning time is generally considered within a second).

Thus we would not expect much appreciation even for simple tunes. Nevertheless tests have shown that monkeys have a small degree of simple-tune recognition.

Now a good musical composition, or just a simple tune, is a whole quantum state in higher dimensions, that is, there is no time in the basic aesthetics of the music at that level. In our 3D, music might be described as art, one bit at a time. However, when we listen to the music with understanding, that span of attention—a single quantum state of energy—constantly references the whole in the higher dimensions in a holographic manner. That is, the part or interval spanned, links to the whole, and the part moves through our clock time but reflects the essence of the whole—which occurs beyond normal 3D awareness (note: education does not teach higher dimensions).

Thus listeners think they are just hearing changing sounds and this is music. However, their attention is spread over a short time interval and this subjective interval moves through (clock/objective) time within a deeper dimension of consciousness. But it is constantly referencing the whole music holographically in the higher dimensions of consciousness's higher aspects.

We see then that rhythm provides a regular repetition, enabling a degree of persistence in time to hold together a sufficient number of sounds and to experience those sounds with a musical and aesthetic meaning. This is rhythm used optimally to assist appreciation of music. But can it be misused?

Inevitably as society drifts into more popular forms of music with greater freedom of expression, the listener will eventually become cognisant that a strong beat with its hypnotic effect assists appreciation where it is lacking. That is, a person may have low attention span for music (which is probably synonymous with lack of musical ability) but by strengthening the beat they are able to grasp the musical essence more easily.

The beat thus is being used as a prop system to hold together the music. The next step is that one experiments with still more percussive rhythms and over a significant period of time, owing to low inherent musical appreciation, one becomes so used to the beat and its advantages that the beat begins to dominate. The eventual outcome of this is that the beat begins to take over the musical essence—the mind is becoming programmed by a past of many recordings of strong beat effects associated, stimulus-response wise, with musical pleasure. And the beat itself becomes synonymous with music for the subconscious mind.

This is now an inversion of the purpose of rhythm and the end result of this 'progression' is that it gradually removes any musical effects altogether. One may see that gradient of deterioration over the years. We now have programmes on television, even documentaries on science and engineering, accompanied by background 'music' that has been reduced to a nonsensical variety of percussive beats—with loss of perception of the original purpose of music in films. (Note that this has nothing to do with, and is no slight on most jazz group rhythms, or a drummer who performs a solo, demonstrating skill and a variety of rhythms.)

Thus we see again another example of science not recognising true unity, only simulated unity. As a result of these scientific limitations and the civilisation's increasing obsession with the quantitative there is a corresponding degeneration in understanding art—it is simply emphasising more the quantitative. There is nothing wrong in experimenting with different forms of expressions but the harmonics and unity of these less artistic or non-artistic forms will be less valuable to the society's enlightenment and evolution.

We can see immediately that correct scientific knowledge applied to art has an important roll here—it is capable of explaining and correcting paths not aligned with true (spiritual, higher fractal levels) evolution. Science could do this in theory even if we are not yet capable of demonstrating this with scientific instruments (such as the oscilloscope mentioned above).

The human civilisation is significantly handicapped in art appreciation owing to organised methods of development of left-brain abilities and discouraging the right brain use. With this imbalance, non-aesthetic forms of communication are infiltrating the art world, threatening the artistic integrity of our culture necessary to our survival as an evolving species. This is not helped by our narrow-minded determination that the Intelligence Quotient

(IQ) test evaluation is the only worthwhile test of intelligence. This will be clarified in Section 18.

Let us outline the principal problem before delving deeper into this abstruse subject. Genuine art today is being replaced to some degree—also infiltrated and adulterated—by *other* forms of communication, which utilise messages, intellect, quantitative aspects of life (as opposed to qualitative) and, in particular, thinking rather than looking. This deviation from aesthetics has essentially risen from the adult's loss of that 'freshness of vision' of the child, envied by the artist; perception is being subjugated by thinking (by 'thinking' we don't mean emotional or intuitive states). Recall the section on creative awareness and automatic awareness. As we age, more recordings are made, causing the automatic awareness (the 'robot' side of the human) to dominate over the creative awareness, which means only quantitative evaluations occur.

What makes art even more complex is that everything which can be potentially experienced in art is dependent on the viewers' understanding, which we can consider would range from ideal to non-ideal, regarding their perception, thinking, experience, judgement, etc. This is a separate variable—the human factor. What if very good art was not recognised, understood, by anyone. What does this mean? We are supposing the viewer may have no artistic appreciation whatsoever. Thus we must assume in our evaluations that we have the ideal situation. If, for example, we are explaining the value of Impressionism then we are assuming that the spectator, or a sufficient number, is capable of comprehending this, which is not necessarily what occurs in practice. Thus clearly the viewer's role plays a vital part in the discussion of the understanding of art, whether it is ideal or non-ideal. Let us firstly take a look at the perception process.

This brings in a further factor regarding the viewer's state of mind, which is that although the universe is organised fractally, the perception process is also fractal, contrary to superficial observation of this process. Attention on (consciousness of) a subject occurs in separate (in time) and repeated whole states (quantum states) or focuses, rapidly switching on and off, and 'superimposed'. [* Similar to the meaning of 'gestalt' used in experimental psychology.]

What this means then is that the perception process itself will tend to group visual elements when they are in related positions. A painter should paint in a manner to trick the perception into apprehending groups within groups of whole states by careful positioning of lines and mass, giving rise to the main attributes: completeness and integrity (wholeness).

The negative side of the perception process is that education develops left-brain consciousness and discourages right-brain development. The left brain will analyse and make comparisons and judgements since it is context-dependent. For example, if a figure within the picture is realistic, the left brain will look at stored images in the mind relating to this figure and make a comparison—this has nothing to do with art. Art is experiential, direct viewing of the pictorial content (what is on the surface of the canvass),* plus experiencing any aesthetic emotional language being conveyed by the subject; it is not thinking—which is representational, that is, looking at what it represents. Do not let a genuine artist hear you say, What does it represent?—when viewing his work. [* This is why the child has an ideal initial state of mind for art appreciation, that is, one deficient in recordings that cause automatic awareness.]

This 'logical' perception involves a kind of analysis, comparison, judgement. It is intellectual, involving the thinking process. The right-brain consciousness understands by direct duplication—there is no 'space' between observation and apprehension in the perception process. It is instant both in space and time. Any delay once the process has begun is a measure of the extent to which the left brain is being used with its detached and fragmented thinking. Or of course it could mean simply that the work of art itself is fragmented, lacking wholeness, and to that degree isn't art—and there is nothing the right brain can do to remedy this. For example, a spectator may view a house in a painting and conclude that it reminds him or her of a haunted house (or, this association may be unconscious). Thus he or she will experience the emotions of these memories, which has nothing to do with art. They may be unaware of this, either fully or marginally, and simply decide they don't like the painting or that in fact it is not good.

The child's consciousness is much more directly involved with what is being viewed. There is that freshness of vision so envied by the artist. Thus when children look at something their attention is on what they are seeing, whereas in adulthood this external attention may only be about 10 percent, and much of consciousness is involved with thinking, with associations, what the object, painting, etc. represents. This is a result of disabling the right-brain consciousness, the in-phase resonant and direct mode of knowing.

Certain forms of modern art have almost established a trend in which artists intuitively demonstrate cleverly this deficiency in the public's proper observation with an indication of its remedy, for example, Picabia's collage, '*Centimeter*'. In this work, the artist has coiled a tape measure clearly to look like a tree, and stuck it in place in the picture. The important point is that he didn't name it a 'tree' but more what it actually is. In effect, the spectator is encouraged to see the work as it is—that is, a tape measure in a particular shape, and not to see it as a tree, in which case one will merely think about trees.

This judgement and comparison (in thinking about trees), will activate left-brain thinking, causing a failure to look at what is actually there. One is supposed to recognise that the tree was merely being used as a framework or vehicle and the art is present in the (maybe) fascinating way the tape has been twisted and manipulated. Note that we are not saying this is good art; it is merely an example of how one should be looking at a work of art.

Other examples of artists 'tricking' the mind to break down its formatting, that is, avoid activating preformatted patterns in the mind and preformatted perceptions, are paintings such as in the impressionist group, for instance, Cezanne's *Still Life with Fruit Basket*. This unconventional, nonlogical design of juxtapositioned fruits and bowls, out of perspective, and nonlinear lines, helps to prevent set patterns from being triggered off in the mind that cause the attention to focus on these within the mind. Thus the viewer must then only look at the painting, and creatively assemble the parts anew. This helps to bypass the automaticities of habitual thought processes. In particular, it aids integration of the parts into the whole picture rather than viewing a 'framed' subject set out and separate from the background, that is, so that the whole surface is the picture and 'subject', as intended. Unfortunately in the example given there is no colour.

Cezanne – Still life with fruit basket

Much of modern art is over-emphasising other forms of communication—meaning they are not art. That is, the other form of communication (such as an intellectual message) is not art, though the rest of the picture or sculpture may be. As long as this is recognised and prevented from pushing out true art there is no problem. Regrettably, however, one of the 'tools' for suppressing man's development and true evolution (into higher frequencies/-intelligence) is this very mechanism of reprogramming the masses with new forms of communication that are not art (unless one wishes to redefine art), much of which is harmless though in itself. But it is educating the quantitative 'robot' side of the human being and discouraging the harmony, beauty, perfection/completion and wholeness of real art; that which can resonate a civilisation into higher states of consciousness.

This is all leading up to the eventual 'cyborging' of the human race (this is not a conspiracy but available information for the masses) in 20-30 years time.[3] Non-art forms will resonate with the lower third-dimensional spectrum, dealing with the reality of excess logic, fragmentation, disharmony, and subjects of conflict and suffering; manifestations of the times (like fashion) rather than be eternal.

Are all messages incapable of integrating as one whole with the pictorial and artistic expression of a painting? No, for example, Van Gogh's emotional messages are not separate, they are an intrinsic part of the aesthetics of the painting, since the feelings invoked, range in the higher-emotional band, aesthetics, and not in the lower feelings, such as grief or melancholy, which are separated away by the illustrative aspect.

Of course, with or without a separate message (intellectual), the pictorial content may not be good art. If it does not communicate beauty, harmony, completion and integrity (undivided wholeness) it is not good art to that degree. It will be fragmented and the conscious serial/linear mind will then merely link together the elements of the picture and seek realistic recognition; and there is no art for the 'unconscious' to grasp. This would now be a composite whole, and thus this whole does not have its own vibration—does not have true unity. As stated, art regenerates an extra dimension of frequencies beyond the 3D 'surface' within the viewer's mind—just as the laser does, described above. Quality is real (a frequency pattern); it is not imaginary or insubstantial.

What about representational and realistic art? Even if it is good art it may automatically cause one's perception to link together the familiar elements of the picture into total realistic recognition, but the associations in the mind of the observer will only create a composite of parts, an image of what it represents, a landscape or whatever—nothing more than a group of recognisable objects, etc. A realistic picture of a tree will invoke the perception process to recognise 'tree', and we may see how realism can (but needn't) distract one from art appreciation—one simply thinks about trees and how well the picture looks like a tree. This was handled by the reactionary approach of the Impressionists. Sufficient distortions of reality were introduced to enable more probabilities and dimensional aspects to be regenerated by the observer. An impressionist painting is sufficiently undefined to bring together extra-dimensional relationships, prompted by the artist's ingenuity in placement and choice of colour, mass, line and brush stroke. Relationships in these factors can stimulate consciousness's aesthetic sensibilities, invoking whole feeling tones elicited from higher frequencies (shorter wavelengths capable of greater refinement) and greater holism. Impressionism demanded, however, creativity on the part of the viewer, who in general is not an artist, and the Impressionist artist was initially opposed and discredited. Today it is the most accepted form of modern art and may be more popular than realism amongst the educated masses.

Realistic art tends to be received as, what we can call, informational art, since all objects within it are copies or are representational—they are things we all have a lot of information on. But for good realistic art it must also have qualitative pictorial content—not just realism. For example, Corot was criticised by the art authorities in that he was being true to nature and not reforming nature as the classical artists did of that era. The academic rules were that one added brown to all greens of nature and every twig or leaf is carefully ordered. What they failed to recognise was that in fact Corot was reforming nature qualitatively, meaning that his relationships within the picture were qualitatively interrelated and organised. As an aside and *ironically*, the pedantic classical paintings of that period—of, for example, perfect trees—resembled a hint of primitive art; highly acceptable as a modern art form today but would have rendered insult to these great works of that time.

The qualitative interrelationship is obtained through harmony of colour, line and masses and good overall composition and design, and a consistency of technical rendering. How much of this would be experienced by spectators would depend on their artistic appreciation. Thus we begin to see the complexities as we recognise not just the presence of pictorial content in art but also informational content.

What can we say about art today? We might ask: Does contemporary art invoke pleasure and a recognition of beauty, etc? Should it? What do we really mean by art and what is wrong with other forms of communication, if in fact there is anything wrong? There is no question that we do see the presence of pictorial beauty, and harmony; we receive enjoyment from its viewing; we perceive integrity and wholeness and sense of completion. But we also see ugliness and many other forms of communication, and a great deal of informational art, most of which contains the emotionalisms of aberration and mere intellectual messages.

This is the respective presence of 'relative' and 'absolute'—in this case, in art. An example of 'relative' is fashion, which one becomes used to and can like or dislike, and it pertains to the times. Nevertheless even fashion could contain some aesthetic and 'absolute' features, that is, they are lasting and have certain basic harmonies that are not dependent on, say, getting used to them.

Much of modern art relates to the times or the current psychology of the artist, and thus tends to introduce other forms of communication, that is, energies invoked by contemporary circumstances, than pure art/aesthetics. It thus will have a tendency to be informational 'art'. However, in addition, it could still contain good art, and ideally should, if it is to be labelled art.

After studying the nature of the qualitative and quantitative one should be more able to analyse human behaviour, art and music, technologies and life in general, and more easily recognise the extent to which quality is reducing in society today in countless different forms. For instance, in human behaviour this is why we have the expression 'good vibrations'. Here we are giving the basics—which is how things work; only physics can provide this at the most basic level.

Does art have to contain perfection? There should be perfection in the sense of completion—nothing need be added or subtracted. But there would be varied levels of complexity of this perfection. An abstract work containing a few coloured geometric shapes could have its own simple perfection. A more complex work would make it more difficult to achieve perfection, to interrelate all parts so that there is no sense of absence or excess. But as we add recognisable realistic subjects, such as a figure, it may distract the inexperienced spectators from the art within the pictorial content by causing them to think about the figure, which is attention on information; it is not part of the aesthetics of the art.

True art is an instantaneous language—not a product of the thinking process. It is experiential; recall that the experiential is a property of the non-quantifiable Absolute. It is not meant to be interpreted (intellectually). It is what it is. Beauty is a language in itself. It is not a mystery to hopefully be explained; and it may never be explained—it wasn't intended to be this. The problem is that the human is not in communication with his or her feelings (in particular aesthetics), and as indicated, this is a product of the Absolute not the mind/robot side of the human; see Section 12. These characteristics have been discouraged and suppressed so often. We all think much more like a computer or machine. Using the model of the human in Section 12, we see that the robot/-intellectual side relates to the quantitative (the parts) and informational art, whereas the input consciousness (interfacing with the 'robot') gives us the creative and unity aspects.

There are different degrees of the ratio art/non-art in most works. The two overlap generally (meaning it can be good art but also be illustrative non-art) but with complete art at one end of the scale and non-art at the other. As we move along this gradient art to non-art, art loses its precedence and non-art begins to utilise art for its own expression. In this broad category of non-art or other forms of communication, art is being used to convey other forms. That is, a message (non-art form) is being communicated and the art form is being used to do this—thus bringing down feelings from the aesthetic sense of true art to lower emotions. In terms of physics, this scale has increasing frequencies and integration (going 'up'). Coming 'down' into lower feelings of emotions the frequencies are lower and there is greater fragmentation of energies (in terms of evolutionary energies, this is 'devolution' as per the New Science). As we move into the informational category we get a picture *of* something—not just a picture. This is what we are calling informational art. It has a subject or subject material. Nothing wrong with this, but the informational aspect is being viewed as art. For example, realism clearly has a subject (or subjects) but these subjects depict the known environment and thus there is only the pictorial treatment of colour, composition, etc., to provide the aesthetic components. The informational content in realism, such as illustrating a sad or gay person is not art. It might be good technique. The illustration aspect may only be a 'copy', similar to the music example of creating weather 'sound' effects, which we shall shortly exemplify.

As we move into distortions of realism, ranging from Impressionism to greater abstractions, liberties are thus taken by the artist, opening the possibilities for creativity, for creating effects which trigger feelings from hidden resonances within the unconscious; what we call a true artistic effect. However, these can range from negative emotions (negative experiences) to positive aesthetics, and the true art value may be compromised.

We have a gradient, or scale, of feelings from low-frequency mis-emotion through positive emotions to intuition and aesthetics. As we come down this scale of aesthetics to emotions, mis-emotions, negative emotions, the communication goes from the natural language of art, such as colour, composition, geometry, which may resonate with unconscious aesthetic structures, gradually to information, relating to our 3D existence, entering into the painting, such as paying attention to the subject, quantitatively, which is inherently detached from art and merely providing subject content and a framework. Its value is lessened as an art form—art merging into information.

The communication might, however, be successful, meaning the artist has succeeded in what was intended. The key is the emotional level and this is where physics comes to the rescue as we have previously implied. We may see the usefulness of physics in clarifying what is happening in art.

Communication of fine art involves a higher frequency than, say, verbal language. The greater the integrity of the art work the higher the frequency—this would be the upper band of emotion and feelings which we are calling aesthetics. Negative emotions, agitation, fear, etc., would be in the lower band.

What can we say about aesthetics? Just as languages have words or unit meanings, which can be combined to form a variety of larger whole meanings (a phrase or sentence), aesthetics will consist of quantum states of feeling tones of particular frequency patterns, which can be combined to form further aesthetic meanings. A state of joy can be a causative, energy-frequency pattern in its own right—not just an effect of something from some external experience. We may search forever for the meaning of beauty, which is basically experiential, but never find it from intellect since it is a language in itself and requires no further breakdown—but it will have a frequency pattern. Our unconscious senses this but leaves our conscious mind with a mystery; like waking up after an abstract dream that was fully understood in the dream state, but now incomprehensible and forgotten in the waking state of linear thought. Great paintings come into being from the unconscious, as stated by many great artists.

These intuitive/feeling/aesthetic states evoke emotions, in fact use emotions but not the lower-frequency negative emotions of, for example, sadness and grief, etc. A painting may contain sadness realistically illustrated, but if it is good art an artistic spectator will not tune into the sadness vibration but the higher-aesthetic feelings. However, another less appreciative person will resonate with the sadness. The latter is not what the painting is about if it has merit—one is expected to override the realistic subject-material influence, involving lower emotions; just merely observe it.

To what extent can informational art be art? Let us consider an analogy in music, which is less complex, and more clear cut. Consider Grieg's piano work *To Spring*. The composition conveys the impression of weather sound-effects: the breaking of a storm, thunder, showers and sunshine, ending with the final pitter-patter of rain drops. Although the impression is excellent, one needn't pay any attention to the impressionistic aspects, or even be aware of them, to appreciate the music. The music is not at all dependent on recognising what it represents. Thus the aesthetics is not sacrificed for the weather impression. The finer waveform (higher frequencies and greater integrity) of the aesthetic concept of the music is complete in itself but it carries the harmonics of the cruder waveform of the impressionistic material. The two, aesthetics and impression, must unify as one whole—which they do.

Now, to represent weather by musical sound requires great skill. In art, however, it is easier to represent such ideas (informational art), and subsequently neglect the art content of a work. The nature of art obviously has pictorial format. It is visual and an idea more easily avails itself to representational information. Thus mediocre artists can express ideas in this visual mode (and become known). But mediocre composers can't express many impressionistic effects very successfully and it is not particularly popular anyway, as a purpose in itself. In good Impressionist art, aesthetics and information (the recognisable impressions) merge because the subject itself is totally conducive to artistic rendering. The nature of the information is based on visual idea (such as an old chair) as opposed to thought idea (illustrating an activity, which is detached from the aesthetics).

Much of modern art, which 'tells' us what to think, depicts the world that we live in. Ideas of the times lend themselves to visual, figurative and informational representation for the simple fact that such ideas are usually visual. However, ideas (of the times) as expressed as sound (for example, music) are not easily expressed. One of the drives behind modern art is to break away from the formal and limited modes of conventional communication, and consequently correspondingly breakdown the extreme formatting of consciousness that takes place in a society such as ours. A person sensitive to this 'narrow-mindedness' of societal programming will rebel against these stifling templates (programming) imposed on our thinking and expression, and the artist may express this rebellion.

Gino Severini – The Boulevard

How important is the presence of subject matter in modern art? Let's consider the painting, Severini's *The Boulevard* (unfortunately no colour). If we envisage it without the subject, the recognisable

Boulevard activities, it still registers as an excellent work. But humans seem to prefer a subject. Yet this poses greater difficulties for the artist. These two factors, preferring a *subject*, plus the challenge of making the subject merge with the aesthetics, seem to enhance the communication and interest. Severini utilises his own cubist style but extremely tastefully and aesthetically. The subject content is evident from the title, *The Boulevard*, and is just recognisable as a boulevard with shoppers amongst the complex but harmoniously and colourfully chosen, interwoven geometric shapes. Thus we can conclude that a subject, even in extreme abstract art, gives enhancement to the work. It gives a more interesting experience and acts as a focus for the viewer to integrate the picture more three dimensionally instead of two (a flat surface). In addition, the artist has also intended this.

Let us comment on technique. It is a measure of how well one is able to represent an idea—it is the skill aspect, and does not give quality in itself. The aesthetic ability of the artist organises the graphical elements to produce a single or unified qualitative feeling, which should of course be pleasurable otherwise it cannot be art, by its own definition. What if the technique is so good that the person could copy a great painting, including the quality and aesthetics; or similarly a musician could copy the 'interpretation', musical ability, of a great musician.

In the case of the musician there is no way even the ultimate level of skill could copy the refinements that a good musician puts into the performance of a piece of music.* The reason is that the refinements are dictated by the whole perception, which is a property of the artistic and musical ability and manifests from the 'unconscious' (higher soul fractal-level of consciousness). Obviously a record can copy it, or a photograph can copy a great painting. The main point is that copying on the part of the artist or musician is only technique and this did not create anything. [* Unless humans were totally robotic, as science thinks, in which case the brain of the robot could make a copy as does a CD but it would take miraculous unimaginable robotic complexities to duplicate this via a muscular system.]

Let us look at the extent to which the artist is dependent on the viewer's perception and appreciation of art. The over left-brain development through our educational system is the first problem. The left-brain analytical mind will not recognise artistic qualities; it will only perceive the parts, and furthermore such a perception will immediately trigger associations of anything pictorial in the artwork. In general, upbringing and education removes that freshness of vision of the child; the ability to actually see what is in front of it.

When the spectator views a painting, if there is recognition of something familiar, we are calling this 'something', informational. The problem is, as already indicated, when there is recognition, the viewer tends to focus on the familiar forms of realism. Even if it is a realistic work the attention shouldn't be on 'how skilful a representation is the object?' These are habitual programmed effects which the viewer must overcome. A good artist when viewing an art work is especially able to break down programming and perceive something as it is, instead of dwelling on unrelated associations in the mind, as generally occurs with the layman. This is the creative awareness and not the automatic (the robot aspect of the human). The viewer may also of course only observe the lower emotional content of the work, assuming it is there; or ideally the viewer may experience the true aesthetics of the art—if it is present.

Thus the artist, when expressing these deep, subjective feelings relating to problems, will (unwittingly) understand their work much better than the viewer. The artist has the background, the unconscious context of what is being expressed; the work can become over subjective, but the viewer only has what is on the canvass.

The number-one English artist Francis Bacon was an interesting and intelligent character. His gruesome paintings were not seen as such by himself. It was his way of re-inventing reality, of breaking through conventional representation with the intention of simplifying and intensifying this reality. He discredited pictorial art and pure abstract as decorative, except where, in a few cases of other great artists, in his opinion, he recognised something more was being expressed. He was fascinated by flesh and meat (as in butcher's meat); also open mouths—gums, teeth, etc. (thus his figures weren't really screaming from his point of view).

Bacon was an excellent artist but his strong themes of apparent suffering and distortions of form will not communicate art or aesthetics to the viewer and will tend to deflect the viewer's attention from any artistic renderings that he was fully capable of achieving—his painting skills and sense of colouring were first rate. However, many of his distorted self-portraits (and other portraits) might be considered more interesting than if they were realistic. They do tend to communicate abstract emotion, which directly correlates with the pictorial content (not as information); a requirement of good art. All we are saying is that they will not attain the highest level that art is capable of. The scientific information is that when we bring into phase, into harmony, previously separate elements by matching frequencies (for example, paint-work or musical notes), a higher-order state is regenerated, that is, a higher spectrum of frequencies, of wholeness of intelligence (from higher-dimensional strata/realms—quantum regeneration). This means the unity (or quality) becomes (is) a 'window' allowing into the third dimension greater harmonic structures/-concepts that couldn't exist in a 3D particle reality only. These harmonic and integrated characteristics of art, as already explained, generate a very positive evolutionary path of the species.

Thus we see that when artists correlate the paint elements—brush strokes, colour, line, mass and shade—the created quality is information coming through from higher-dimensional realms (higher spectrums of greater frequency rates). This is also what 'spiritual' means. Science will eventually recognise the existence of these higher fractal levels of consciousness. Everything apparently is fractalised.[4]

One of the mechanisms that gives an outlet for quality is fractals: an expression of degrees of order and a genuine form of art—see earlier sections. It is not a subject the layman will welcome in its technical aspects but good artists have been using fractals expertly since the incipience of art in society. Fractals is a key item in integration and quality. Each system of fractals contains a repeated self-similar pattern but of changing dimensions. If one gazes at the unity of the form of a fern (perception of the whole) one finds that unity repeats in smaller scales in its leaves. A work of art, with a recognisable subject, depending on its complexity, should look balanced at several levels of perception, experienced at different distances from, say, a painting. The details within the picture should be grouped but as one steps away using a wider focus the balance of mass, light and dark through correct grouping, should still be present.

Consciousness is an intrinsic essence of the Absolute and the infinite fractal is inherent within consciousness. When it resonates with the environment we can experience beauty, art, etc. It can resonate with its own higher nature and perceive qualitative effects accordingly. That is, a lower fractal level of our consciousness resonates with our higher fractal levels (soul, etc.) providing we are creating qualities. Note that consciousness in higher fractal levels will have additional senses.

Fractals are often exploited in art forms such as Pollock's action paintings, which are skilful fractal designs. Thus use and application of fractals will aid integration and the communication of quality. Quality acts like a window into higher dimensions of the mind and spirit (this is the inner-space or inner-consciousness). Or we could consider it a window into the unconscious. Good semi-realistic art, such as Impressionism may be less defined but sufficiently integrated to enable consciousness to 'fill in' the gaps (the windows) so to speak with other probabilities and interrelationships; this is where the soul energies enter. Realism limits this; every little point tells exactly what to see third dimensionally. Note that the soul level is merely our next fractal 'up', from which we were 'projected' as human extensions similar to the way a branch projects (grows) a twig.[5]

The development of human civilisation has stressed logic rather than feeling. This suppression of natural and soul-level feelings and failure to allow the subjective and psychological outlet causes enforced expression—a type of learning by misadventure. As a result, wild, chaotic and materialistic expressions ensue appropriate to this suppressed, confused state of the unconscious. If higher intuition is repressed, the lower emotion may react into expression. Thus artists bring through a wide variety of emotions. If we were all highly sane and evolved, nearly all art would relate more to harmonious subjects.

Impressionism is probably the purest form of art, plus some modern works, such as abstracts that have aesthetics as dominant. But good art does not have dominant messages, with which some artists get carried away. In good art, a landscape, for example, is not for communicating primarily the scene as realistic information—it is communicating a qualitative view of the scene. The artist is reforming nature qualitatively. It isn't necessarily there in nature but can be triggered within the soul of the artist.

As explained, by soul we mean a higher fractal and wider focus of consciousness embracing greater degrees of freedom and order, of which our conscious state is but a 'hypnotised' (relative to 3D) aspect of this greater state. The artist simplifies and extracts from the view or any experience the principal attributes, and will sometimes accentuate them; then puts them together as a whole. Thus these larger qualitative wholes are soul attributes/states and can be integrated into forming a work of art. An attribute may be simply (the quality of) 'treeness' or snowness', etc. Some people remember as a child, experiencing qualities of ordinary things. A realistic picture may have too many bits and pieces, and be unrelated, which either do not express the quality, or distract the viewer from the qualitative attributes when present.

Quality from the soul, although can be expressed in 3D, can't be logically detected and defined. We can only hint at or suggest what is happening. Just as we have clear-cut emotions, at least we identify them as such. The soul level of consciousness will have attributes of feelings which in terms of energies will be frequency 'packets', such as joy, gratitude, etc. To us, many of these will be abstract and manifest somewhat weakly in 3D for most people.

The harmony and aesthetics is far greater at the soul level though. The soul level downloads its 'feeling tones' (aesthetics, harmonious frequency states), which to us in 3D are abstract higher-emotion attributes (higher-frequency coherent states/-concepts). This aesthetic rendering, of what would otherwise be, say, merely a photographic view of the environment, enhances enjoyment of the art work, triggering those soul sensations within the viewer. It is as though these feeling tones (frequency patterns) are the 'building blocks' of the soul/spiritual body and may be called archetypes. Higher aspects of consciousness, such as the soul, over-soul, etc., have higher-frequency senses.

The artist is creating a new context for or within 3D by bringing in these higher values and integrations. These states of quality are duplicating the native, innate properties of the spiritual being's modes of consciousness. Good art will resonate with the fundamental and archetypal states of the soul.

Let us briefly look at the matter of harmonious subjects and nonharmonious subjects. As we have implied, subjects are illustrative, but this is not art and must never be the sole purpose of art; just as impressions (recall the example given in music) must not be the purpose; there must be sufficient aesthetics apart from the impressions. Much modern art though has moved away from this recognition, and negative subjects are being brought through for their own sake. If the art rendering of the subject is good though, then we may still accept it as the 'fashion' of our times; as a contemporary expression of reality on planet Earth.

A very different factor influencing the art world is programming, or mind manipulation, to embed contextual biases within the unconscious. By these techniques an artist can be selected and promoted to give a very positive status. A good instance of this in the field of science would be Einstein (now a household name*). This type of brainwashing can hugely mould minds towards excessive favouritism of certain artists or their works. Vice versa, specifically in the field of knowledge and politics, background contexts can be built up by this programming—particularly as groups form, supporting a particular view to create opposition. This comes under covert politics and will not be pursued.[6]

[*Regarding this example of Einstein, such emphasised selection can for instance cause a by-product of that person to embed in the mind of the population, or scientific community, such as relativity—and it becomes very difficult to overcome the resistance to upgrade or question this law.]

Finally, what would be a guide line as to how to identify what is genuine or good art? An art work to be called art must be defined as possessing the following qualities as a minimum: it must have unity and harmony; it must be pleasing; it should have a sense of completion—one cannot add or subtract from it without upsetting the balance and integration (as far as possible for the particular work). In terms of physics this is a single quantum state of perception (frequency pattern) and when it occurs is instantaneous—beyond space and time. These qualifications could now be called 'aesthetics'. Thus the primary goal of any art must include at least these qualities.

Notes
1. Eventually the scalar-electromagnetic oscilloscope will be developed which will pick up the high-frequency scalar energies of the mind and reveal each person's identity in the form of a spectral analysis. Even great art could be evaluated in this way. Chapter 18, *The Original Great Pyramid and Future Science*.
2. Ibid.
3. Book: *Engaging the Extraterrestrials: Forbidden History of ET Events, Programs and Agendas* by N. Huntley.
4. Booklet: *The Fractal Tree*. www.nhbeyondduality.org.uk.
5. Ibid.
6. Op. Cit. *Engaging the Extraterrestrials*.

17.

THE NEW SCIENCE IN SPORTS AND ATHLETICS

Modern physical training for sports and athletics, in fact all skills, is not complete without an understanding of the learning pattern (mind computer programs) which plays a major role in muscular activity.

We shall use the human/robot model and its input function or soul, described earlier. Let us look at some of the fantastic mechanisms of this structure, the 'robot'. As mentioned, the structure or robot side of the human is very much emphasised in our civilisation/evolution, which is quantitative and is measurable, whereas the consciousness or soul component possesses absolute and non-quantifiable characteristics. But keep in mind it also has higher-order mind structures, which are potentially quantifiable. Recall the non-quantifiables, such as the 'aliveness' property, the sentience and the experiential.

Even in the performance of musical instruments the technique/skill aspect is entirely the subject of robotics, that is, computer programs (and therefore it is the property of the 'robot' side), and thus the ability is based on practice and the build up of programs or learning patterns. But the musical expression is achieved from the 'soul' aspect, as is intuition. These latter characteristics function in the present moment.

Let us consider the mechanics of physiological motion of the human (robot) using muscles. This is achieved by means of the learning pattern, already mentioned. The learning pattern provides control over physical movement; it is a 4D holographic template, where the programs are stored. Practice of movements enables coordination within the learning pattern to occur but also expansion of the learning pattern is required. Thus the learning pattern is the basis of all skill ability in all activities, such as sports, athletics, instrumental technique and skill generally. Let us explain this.

Readers and students of this new science of physical training will be surprised to find that the mechanism of physical mobility is far more sophisticated than their education has taught them. Current science assumes that physical movements of the body/limbs of humans and animals are entirely governed by the central nervous system (and autonomic nervous system), in which nerve impulses are sent to the muscles, causing contraction and subsequent motion of the body parts.

If this were the only mechanism present, the limbs would do little more than twitch; there would be no fine coordination, no control, and no ability to access the system (keep in control of the movements) while motion is occurring. The author, during 50 years of research has discovered that physical mobility (voluntary) is due to a dual system: 1) the central nervous system, and 2) an energy-field system. These two systems are synchronised perfectly. This energy-field system is a type of computer mechanism and is in fact nature's quantum computer system, an understanding of which is much sought within most scientific disciplines. We are normally only aware of sensing muscular activity, but if we, say, *imagine* moving the arm (no actual muscular activity), we will feel kinaesthetically the motion. This is the sensation of information within nature's computer system. (The neurophysiologist will tell you it is the physiological kinaesthetic sense around the muscles and joints, but in fact this accounts for only a tiny fraction of the overall sense.)

What must be recognised then is that there are two entirely different systems of training for these mechanisms of body motion: (1) The brain/body mechanism, and (2) what we can call the quantum-computer system.

Everyone is familiar with physical training, which comes under (1). There is adequate knowledge and methods available for this activity in current physical education, and we shall not pursue this field. The other mechanism, nature's computer system, consists of an immense complex of energy fields (quantum/scalar/electromagnetic fields) around and within the body and limbs, which are, of course, invisible to the naked eye, and not detectable by existing scientific instruments, which can't handle higher-frequency electromagnetic waves or scalar waves. This system requires a very different kind of training, which is essentially the development of the learning pattern.

The two mechanisms involved in physical activity, that is the muscular and the energy-field system are very different from one another, but nevertheless they combine and interface successfully. The muscular system provides the basic force and obeys Newton's laws (of forces, inertia, mass, acceleration). However, it is a non-harmonic technology;[1] a force system, whereas the quantum (energy) field system provides a harmonic technology, a non-force system that bypasses Newton's laws (it will be sometime before scientists believe or find out this). The two systems, however, are quite compatible. The field system acts around the joints, and assists the dynamic action of the physical movements, but under high development of the learning pattern, within the energy field, inertial effects of limb movements are reduced. This is part of what occurs in increased skill.

What other kinds of improvement are to be expected from the application of this new knowledge? The muscles will reduce their background tension; effort to make movements will decrease; coordination ability, that is, accuracy and speed of learning, will increase; a sense of improved reflexes will occur; less attention will be required to maintain tensions in the muscles in appropriate activities; speeds of movement will increase; automaticity of complex coordination will increase (less attention required); endurance in repetitive motions will increase, and tensions in the muscles can be increased and decreased more rapidly. These are the main benefits that can be derived from development of the learning pattern.

The educated reader, knowledgable in related fields, would be expected to be sceptical and patience is strongly advised, as the author experienced similar scepticism prior to establishing the material over a period of many years during observations and evaluations.

The attainment of greater physical abilities is governed and influenced by many factors. We can roughly divide these factors into four groups: 1) the physiological, 2) the quantum-field system, which also embraces the communication system between mind and movements, 3) the psychological, and 4) the spiritual.

The first one, the physiological condition, is given the greatest emphasis in our society at the present time. This is to be expected since it is the physical body which is moving and clearly has appropriate mechanics involving physiology and physics, including chemistry and biology. This is well taken care of by training methods today. We do not need to elaborate too much on or reiterate the findings of physical education that are well documented, except as it relates to our main topic, which is the second one, the quantum-field system. This field system controls the nerve impulses in voluntary muscular actions, and can be visualised as existing around the joints. But also it is an additional system that further acts to create physical movement, particularly in high speeds and complex coordination. This provides every detail of every movement.

The third factor, the psychological, has finally been recognised as being significantly influential in aiding athletic abilities, for instance, utilising positive thinking and visualisation, etc. Sports psychologists are playing a more prominent role in aiding sporting and athletic abilities. This factor and the fourth, regarding the spiritual have been dealt with in the author's other writings.[2] Here we are mainly interested in the mechanisms of physical movement.

Our principal subject then is number two above, the quantum-field system, which is a separate mechanism from the muscular and physiological ones. It could nevertheless be regarded as purely physical, since it involves the precise recordings of movements. However, these recordings (storage of information) do not primarily occur in cells, but what we can call the electromagnetic oscillations of quantum stationary states. Thus the information/-program for skills, general movements, etc. is stored in these countless superimposed quantum fields around and within the body. We are not talking about anything metaphysical. The system is purely mechanistic; it is robotic and programmable—within it is structured all learning patterns, habit patterns and programming. Note that the term 'habit pattern' sometimes is used interchangeably with 'learning pattern' but strictly we should use the term 'learning pattern' for movements which are entirely within one's control and are voluntary. We shall use the term 'habit pattern' for learning patterns that have gradually become fixed and inflexible. They tend to hinder the sensitive relation of the cause-and-effect feedback interface between consciousness's control and the learning pattern; a little unconsciousness has crept in, and this latter condition must be addressed to restore the learning pattern. The meaning of programming in physical movements is merely connecting together already learned movements, that is, forming new multiple movements and coordination.

We shall come to understand though that there are two different types of learning process in physical activity. One is specific learning, which is merely programming, that is, coordinating and connecting together movements, and the other, a more general type of learning, not recognised in current science;* simply the ability the make the movements themselves, which is governed by information capacity of the quantum field within the space of the movements. This will be explained shortly. [* Experimental psychology has finally recognised a more general form of learning but not how the mechanism works.]

Recognise then that this quantum computer system has nothing directly to do with the psychological or the spiritual—it is not of general psychology or parapsychology. It does, however, connect with the physiological mechanism, the central nervous system and musculature.

Thus, as already stated, the field of coverage in this section includes all skills, sports and athletics, and all instrumental technique; anything that involves physical body movement, and the subsequent learning process. Current training methods, which are based on physical education and are directed at muscular exercise and development, are not complete; they are not taking into account the knowledge of the communication system from the learning patterns in the mind. However, not all of the above-mentioned activities require physical training; for example, development of technique in musical instruments, or snooker, do not need further physical exercises beyond that obtained by normal practice. These activities involve sheer skill; any dynamic strength that is required is adequately provided by the development of the learning pattern. Thus we are not concerned with the subject of physical training, which today has been thoroughly researched and more or less perfected. Nevertheless all these activities require the use of muscles and the learning pattern. Muscles provide the basic force for movement and the learning pattern regulates the patterns/-programs sent to the muscles. The learning pattern is part of the mind-computer system and contains the programs for all movements (as we have seen earlier, the mind is a higher-order system than the brain).

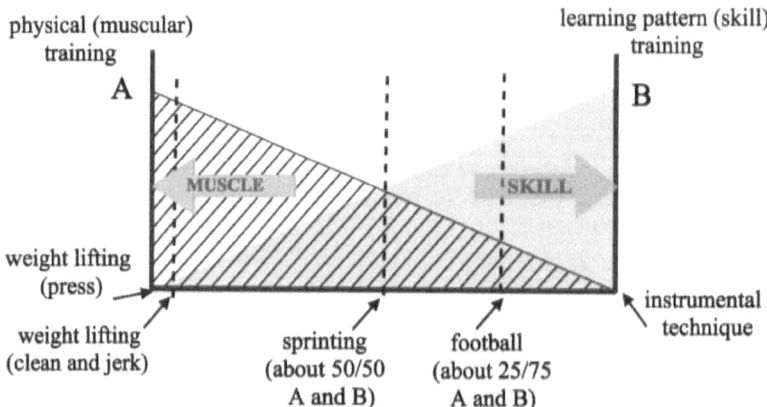

FIGURE 18: Shows the relationship between type of physical ability and what type of training is required.

In this whole field of activity—sports, athletics, instrumental technique, etc.—there are two systems to develop: the muscular, and the communication system to the movements from the learning patterns. What might be helpful here would be a graph inclusive of all physical activities, which will indicate immediately the difference between the activities regarding what should be trained and how much. See Figure 18.

The extreme left of the graph would be all activities that are entirely muscular dependent; sheer strength, no skill. On the right side, muscle training is zero and skill is highest. On the extreme left then, we have the example of weight lifting, but slow lifts. As we consider faster lifts, such as the clean and jerk, an element of skill is introduced, which assists in lifting a little more weight. On the graph, the position of, say, the clean and jerk will be a little to the right, meaning that the opposite side of the graph, representing entirely skill, no physical training required, is being incorporated to a small degree. Instrumental technique is on the extreme right, not requiring specific physical education techniques.

Let us take another sporting or athletic event, say, sprinting. This would be expected to be fairly central in the graph, meaning that the sprinter requires both muscular strength—particularly for acceleration—and the skill factor, of which a by-product is speed. Clearly the act of running requires little skill, meaning the average person can run adequately. This fact gives rise to the omission by the sprinter of doing training for improving skill; not realising that the general learning factor (which raises the level of the specific ability) will speed up the reflexes. This means the sprinter should spend about equal amounts of time on physical training and learning-pattern development. Recall that the specific learning is the actual learning of the particular motions executed in the activity, which involves increased coordination and accuracy, but general learning expands the learning pattern so that the specific can attain a higher standard. However, it is the general learning which ultimately increases speed.

Now let us examine further sporting activities. In Figure 18 if we consider what position on the scale the sport of football would be placed it would probably be around the 75 percent mark: 25 percent physical training and 75 percent skill. Of course a brilliant player has already developed the skill factor to a high degree and might spend all his time on actual performance and physical training and succeed in retaining the ability over the usual expected period of time. Performance will, however, deteriorate if and when there is lack of physical training, whereas the skill factor is more easily retained, and there shouldn't be any loss as long as games are played fairly regularly.

The subject of aging arises here. Since science doesn't recognise the true nature of the learning-pattern (the skill mechanism), it isn't being realised that decline in ability to some degree in most sports and athletics, and a great deal in others, such as snooker, and instrumental performance, is not due to physiological aging. Regarding development and expansion of the learning patterns, the latter can easily turn into habit patterns and no further improvements can occur. This occurs more easily the higher the skill level. As skill (coordination, accuracy and speed of reflexes) reaches a high level, the required amount of tension in the muscle naturally decreases precisely.

This beneficial decrease will be eventually held back by the instinctive belief that more tension, effort and power, is good, which, in fact, it is for immediate results and establishing coordination. But eventually this will cause the learning-pattern template to set like concrete. This occurs with emphasis in professional snooker; the revealing plateau of no more improvement arises far too soon.[3] By applying appropriate training, that plateau can be avoided (or potentially restored). The inevitable physiological one will occur much later in life.

In the simplest possible terms, let us explain how the learning pattern works in performance and development, revealing the type of problems that can arise to arrest development. The subject of skills was the chosen subject by the author for a doctorate in experimental psychology.

Thus the learning pattern is the basis of skill ability and we have two categories for its development: 1) specific learning, and 2) general learning. Experimental psychologists some 40 years ago focussed on specific learning—hence the frequent use of the term, the specificity law. Finally, today a more general learning is recognised as being valuable in assisting the development of the learning pattern. Let us define the learning pattern again: a 4D holographic template that houses programs and converts nonlinear information into linear information.

What exactly are specific and general learning? A simplified description can be used, taken from computer programming. When programming a computer, as the program develops, some of the computer memory is used up. Computer bits provide the information instructions for the program and when the programmer runs out of memory, the computer displays a message on the screen, 'out of memory'. This will relate to the total memory; when it occurs with the human, say, practising some skill, firstly the mind does not inform us of the unavailability of 'bits', and secondly this depletion of memory bits is a common occurrence in practising movements, and the absence of 'bits' will refer to the local regions of learning patterns associated with the required movements (not the total memory).

Now the creation of the program is the specific learning—entirely a matter of connecting up the 'bits', that is, coordination. The process of adding more memory bits gives us 'general learning'. This will increase the density of bits/information, whereas specific learning is simply the process of using up (interconnecting) the bits to form a program. We can simplify this to clarify the importance and difference between specific and general learning.

Take a sheet of paper and pencil. Draw a closed boundary, such as a circle. With the pencil, place dots in the area. This now represents a simple computer memory; the dots are the 'bits'—the memory capacity is the number of bits; see Figure 19. For those who don't know what a computer bit is, just imagine each pencil dot to be a very small unit of electrical charge. When it is activated (given an open pathway) it moves and in combination with other bits, it performs a function, such as sending a signal, command or instruction to achieve a particular task.

We now simulate a small program (A to B) by using the pencil to connect dots as shown in Figure 19. The result demonstrates a program path which could, for example, be a movement of the hand, such as on a keyboard with finger action. When A is activated, the program sequence is triggered and the string of bits is fired off.

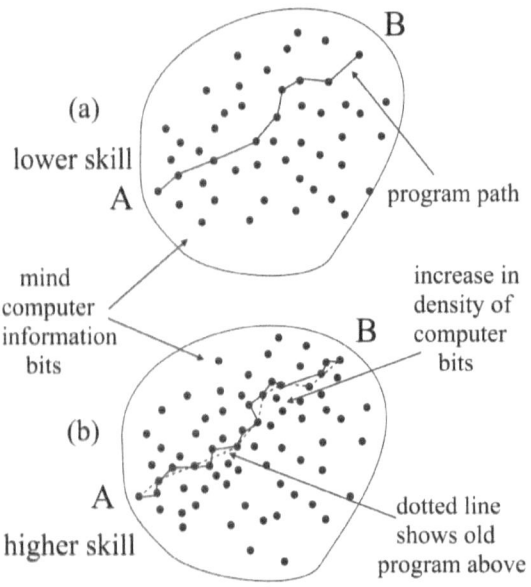

FIGURE 19: 1) Choice of path constitutes specific learning; 2) increase in bits is achieved by general learning.

We can see now that connecting the dots is coordination; this is specific learning. Although the path can be adjusted, say, by reconnecting a line to another dot for greater accuracy, we can see clearly a limit to the development of this program and its subsequent output; such as the finger sequence on a keyboard. This sums up the existing recognition in our education as to what takes place in the process of learning a skilful movement, meaning this is all that current science recognises.

Can the reader now spot what general learning would be? When one practises only for specific learning, one soon reaches a plateau of improvement; the graph increases rapidly initially then curves off into a horizontal with no more improvements. General learning raises this plateau to a higher level of skill—that is, the potential for the specific learning is increased. Thus general learning will increase the density of computer bits (the dots, see Figure 19) so that the specific coordination process has more choices of bits.

In terms of physics how does the density of bits increase? The existing bits in the coordination process undergo fission, that is, multiplication of the number of bits. [Note that there is nothing extraordinary about this; these are minute bits governed by wavelength—small quantum states of scalar-electromagnetic energy—they undergo fission (division) far easier than the more complex process of cell division (billions of times larger) with which we are familiar.]

There is not the scope here for dealing with the general exercises,[4] but briefly one executes rapid light relaxed repetitious movements which are a breakdown of the specific coordinated movements. The emphasis is on relaxation; one reduces tensions in the muscles to a value below the threshold of the required (full) tension of the movement. This encourages the mind to 'fill in the gap' (the missing higher tension).

Thus as skill increases, greater relaxation (less tension) occurs in the muscles, but this absence of tension is replaced by the quantum-field density of information (bits)—this provides extra dynamic power (not specifically static strength). ('Static strength' is generally what we normally define as strength, which is strength in the limit of no motion against a resistance, such as gravity. Dynamic strength is the force generated at speed—defined as 'power'.)

One might insist that the increase in dynamic power when skill is increased is merely due to muscular developments in particular, increase in efficiency in the physiological processes. This would be true, but there is also a field effect which assists the muscles directly.

An example is revealed by the following comparison. Let person A be the strongest man in the world (which is always measured more in the static sense than dynamic), and person B, say, an international (test match) cricketer. The cricketer is a top bowler but also an excellent fielder. His throwing action has reached a high level of skill. We now give A and B a weight to throw, beginning with an appropriately chosen heavy weight. The strongest man, A, throws it the furthest, but as we lighten the weight sufficiently, the cricketer will succeed in throwing it the furthest, such as with a cricket ball. The cricketer has transferred more energy to the ball and demonstrated more power for the light object.

Let us now briefly examine impact skills. This activity is any skill in which some form of impact arises in the sequence of movements, such as golf, tennis, even piano-playing and, in particular, martial arts.

What is the difference between an impact skill and an ordinary skill? Many skills may have an impact feature associated with the movements, even though of a mild nature. Certain points of contact are made and these connections or impacts are utilised by the program or learning patterns. The learning patterns will, where possible, designate these contacts as reference points for the next movement or a reference point for tension adjustment, etc.

Piano technique involves this impact feature. When a finger contacts a key this fixed point acts as a feed-back system, which the learning pattern utilises to perpetuate an accurate sequence of movements. However, what is of principal interest in impact skills is that the moment of impact is very brief, about one thousandth of a second, and this means the final result, such as for instance the ball trajectory in golf is governed by what happens to the ball during that brief moment.

Thus this is the feature of interest in impact skills, which is dominant in such sports as golf or martial arts (breaking the board). When a golf club hits a golf ball there can be an impulsive force of up to 40 lbs. This would have to be much greater for the martial arts adept. However, let us proceed with the golfing example.

This 40 lb force will create a significant jolt at the joints of the body: ankles, knees, hip, waist, shoulders, elbows, wrists, and fingers. The joints will tend to yield during impact. To the degree there is any yield at any joint there will be loss of power transfer.

Conventional knowledge tells us that since impact time is around one-thousandth of a second no change can take place in the musculature during this time. The nervous system cannot react within an interval of one-thousandth of a second. Reaction times are usually around one-eighth to one-tenth of a second, though one-fiftieth of a second is claimed to have been measured for certain muscular processes. Nevertheless, the muscular system cannot remotely act (via the brain/nervous system) within one-thousandth of a second.

The good news now is that the quantum computer/field system *can* react within this time interval. Consider piano technique. A concert pianist may be playing more than ten notes a second but he knows exactly what each finger is doing at every instant, not to mention knowing the state of every other muscular tension. He can intervene at any instant; he has to for proper musical rendering (which is in the present moment; not automatic/robotic).

We are talking about electronic times to instantaneousness. How has science overlooked this; or ignored it? Not only this, but when individuals are executing a continuous sequence of co-ordinated movements, they can take their attention off the movements and return it so that the movements are instantly within conscious control again. This is an astonishing computer access system, which we have already described.

Let us return to the golfer. In terms of current knowledge, the professional golfer, normally well-relaxed in the swing, would have to tense the muscles significantly before impact in order to lock the joints in a manner to prevent yield and loss of power. This clearly invalidates the relaxation cycle, and in fact it does not occur. What we are leading up to then is that the energy-field system around the joints, which operates on quantum-field principles,[5] can act to lock the joints during impact, during that short period of time—about one-thousandth of a second.

When the club head contacts the ball, a reaction of, say, 40 lbs occurs on the body and the joints begin to give way. This sense of resistance against movement, due to the inertia of the ball, is counter to the intended program within the learning patterns to continue the movement unabated. A feed-back system goes into operation to counteract the yield which is trying to occur at the joints. In precise proportion to the resistance, the quantum fields lock the joints to prevent yield. Strictly speaking the quantum fields lock the joints but allow movement to continue without drag—the muscular system couldn't do this. This requires of course a high density of information to have developed at the joints from practice.

Consider a pianistic example. A concert pianist can strike the keys from a position well above the keys in a very relaxed manner such that the fingers are quite relaxed before impact, but will lock instantaneously on contacting the resistance from the key action. As the quantum field's information senses the resistance, causing collapse at the joints it compensates—as per what is intended.

This quantum-field information operates more for high-speed movements. It does not particularly appear to aid strength significantly in slow movements, such as pressing a barbell. More specific research would be required on this but we must admit that generally a naturally athletic person, regarding speed ability in, say, many activities, is usually stronger, as well as faster, than average even in the slow, forceful movements. This possibly indicates that this person with obviously good development of the quantum computer system, since he or she is fast and well coordinated, is also stronger owing to higher-information density. The initial conclusion is that dynamic strength is greater but not necessarily static strength (what we normally refer to as being 'strong').

At least we see that the quantum-field system essentially gives greater power, which is velocity times strength, in addition to skill. We are not saying there will be no muscular tension contributing to locking the joints. The two systems, muscular and quantum computer work together and are perfectly synchronised. But those muscular tensions can't respond during the impact time of one-thousandth of a second. It depends on the level of ability as to how much the quantum computer system is used relative to the muscular system.

One may see now why a professional golfer can hit the ball so far, or the martial arts expert break the board. The energy stored in the swing of the professional golfer is transferred to the ball much more efficiently than for the amateur golfer—that is, it is transferred during impact. It is not just a matter of timing. All the joints must be locked as simultaneously as possible, of course, but this takes skill and practice, and will develop alongside the principal factor. That is, by the time the professional golfer has developed sufficient information density within the learning patterns (the quantum computer system) for successful transfer of energy he or she will be able to lock the joints simultaneously.

In martial arts, imagine the strongest man in the world competing with the expert in breaking the board. The strongest man in the world might be twice as strong (let's say), and yet not succeed as well as the martial arts expert—why is this? Let's say the strongest man generates 100 units of energy in the moving arm and the martial arts expert, 50 units. The impact time is around 1/1000 second. Since the strongest man has not developed the skill and concentration, that 100 units is spread out; not focussed well. Let's say only 20 units of energy is focussed during the 1/1000 of a second to create the force. The expert may focus all of his 50 units into that short time interval.

We can understand now why a stronger man may not be able to break a board with his hand but a weaker one, expert in the martial arts, can succeed—the focus is the key, and the slightest yield in any of the relevant joints can result in considerable loss of power. As soon as the martial arts adept strikes the board, the resistance to motion triggers a feed-back system to compensate by preventing collapse at the joints during the impact time, around one-thousandth of a second. It is quite a miraculous mechanism.

Note that in martial arts, for example, aikido, energy can be created by a definite act of consciousness, or the mind, producing super-normal results in addition to the logical physical effects. One of the names given to this energy is chi.* Nevertheless the automatic mechanism, the quantum computer, can achieve astounding results with considerable training (developing the skill) and this is quite automatic, but it cannot account for any sudden changes in achievement (by applying the mind), or account for the sometimes amazing protection that occurs from injury. [* Like a tenuous plasma; chi and ki are the polarity components of the energy, prana.]

Now let us return to general and specific learning. We have explained that general practice will enable the specific ability (the actual coordinated movement) to reach a higher level. Recall that we stated that speed was also a by-product of skill. Extra speed, such as required by the sprinter is gained particularly by practicing general movements (relaxed, rapid, repetitive movements about the joints relevant to the activity of interest).

Can the reader see that by increasing the density of information in the movements involved in sprinting (see Figure 19). This will enable faster motion of limbs to be achieved, since it means the frequency of impulses will have increased (as the learning pattern expands, its highest frequency increases).

Why can't a fast movement be made by simply applying more force; as current knowledge would tell us? Science has revealed that continuous application of more force in the achievement of high reflex speed meets up with a mysterious resistance to motion. As the speed of, say, a leg or arm motion increases, it eventually becomes limited by rate of information (frequency). The faster motion is endeavouring to 'stretch;' out (in time) the density of the computer bits. The mind computer design does not allow this beyond a point at which the control or coordination of the movement is significantly lessened, caused by this lack of bit-density or control points. The most that would be allowed would be in simple movements not requiring great accuracy, such as the arm action in boxing, or to a lesser degree, even legs in running (since high control is not required).

Today, the champion sprinter can develop a great power output in the leg muscles, but that extra speed will be limited by rate of information. This not only means general training of the immediately applicable muscles will be beneficial but also general training (spatial) involving other muscles of the body. The sprinter, in particular, is using most muscles of the body, and thus needs to train these with general techniques roughly in the order (of degree of activity of the muscles): legs, arms, then various other muscles groups, such as abdominals, that are undergoing tension during running.

Notes
1. Book: *The Emerging New Science*.
2. www.nhbeyondduality.org.uk. Articles on skills and book: *The Attainment of Superior Physical Abilities.*
3. Ibid. Snooker booklet, *For Snooker Players* by N. Huntley.
4. Op. Cit. Book: *The Attainment of Superior Physical Abilities*.
5. Ibid.

18.

THE NATURE OF INTELLIGENCE

Intelligence takes many forms, well beyond the simple standard testing within current education. We also explore the underlying principles of basic intelligence from the qualitative/quantitative relationship to the function/structure viewpoint, and finally the most fundamental system of all: that of orders and organisation, beautifully demonstrated in our fractal, holographic universe of ordered dimensions for the exploration of consciousness.

[Reference: abridged from The New Education series, *The Nature of Intelligence, Part I and II*, by N. Huntley www.nhbeyonduality.org.uk.]

The standard form of intelligence measurement for our society is the IQ (Intelligence Quotient) testing. This is not universal; it is peculiar to our stage of evolution (and deviations from it), that is, it is a by-product of human's evaluation of what is important in the field of abilities. Since we emphasise left brain, intellect, logical ability and have taken an evolutionary path emphasising the quantitative, the IQ ability is very useful but is not the only form of valuable intelligence, which we shall explore shortly, but ultimately the most basic form of intelligence is degree of order, and this would apply to all creation throughout the universe, whether an atom, cell, scientific instrument, planet, universe, person, etc.

This universal attribute in fact has been the essence of much of the subject in this book, in particular, fractals. We shall return to this.

There are clearly many modes of intelligence; however, we might note that probably most forms are contextual, that is, the existence of useful intelligence is a development determined by the application, for example, a skill (we might even consider that the IQ product is contextual). The human may use several modes of intelligence simultaneously—but this can then mean that the combination may be simply a form of intelligence in itself as a whole, and we have analysed it, broken it down into non-useful or lesser useful abilities in our theorising and then discarded them. Varieties of intelligence have been recognised by scientists and educational writers but only those forms have been given attention that are functioning clearly within our society and furthermore limited by existing models of the mind and personality. These forms are also the product of negatively-oriented educational control systems (such as promoting quantitative rather than qualitative talents).

Should we assign a level of intelligence to any ability? Perhaps not. For example, an insect might have phenomenal flying skills but this we know is due to an inborn program we call instinct. Thus creative intelligence should not be required. But if we were to assign intelligence to our computer and robot mechanisms then we must also give intelligence to the insect. At this point we shall simply acknowledge that the origin of this mode of intelligence, based on programs, the collective, nature, God, is highly intelligent, but not the creature. Even so it is not so simple since how do we know the insect's consciousness does not contribute some control, even if small, over the mechanism and actually does occasionally make simple choices (from its own higher aspects).

Intelligence involves behaviour of some kind. This may be mental, physical, and emotional, with biological and genetic influences; intelligence must include programs and learning patterns. We might consider theoretically that raw (native) aware-ness, such as the Absolute—a more pure form of consciousness—to contain potential wisdom and refer to it as general intelligence. However, for our reality purposes this awareness needs to have an application, that is, it needs to be formatted; this then becomes directed or specific intelligence.

Before we attempt to put any order into this broad subject, let us list some general features and expressions of intelligence: the ability to recognise and solve problems, apply logic and act rationally; rate of learning, memory, and the ability to acquire and hold information; ability to evaluate importance; put order into disorder; persistence in achieving a goal in spite of obstacles; consciously hold a focus; ability to imagine (experientially) some new experience which is merely observed in another; recognise differences between harmony and disharmony; self-evaluation and, in particular, the ability to recognise prejudiced perception in oneself; to draw from the intuition; recognise behavioural objective contexts of society that are acceptable; higher sensory perception; abilities in coordinating a body; musical and artistic talents; quantity of relevant data detected in the environment; common sense; animal type of intelligence ranging into psychic intelligence; the ability to step outside closed systems of thinking or programming; and of course we can hardly leave out responsibility.

Let's make a few comments on these before looking for more basic categories. Most of them are fairly obvious and useful abilities. Common sense would tend to relate to being rational most of the time and not be influenced by personal aberrations and prejudices and exercising a natural ethics. These people will not be overprogrammed, are not likely to be dogmatic, and are more capable of giving an unbiased opinion.

Responsibility is fairly obvious as an intelligent trait, but it would also include the ability to confront situations. That is, not to push aside certain unpleasant circumstances in the hope they will go away (the ostrich effect).

Psychic types of ability have not so much taken time to appreciate as they have been suppressed. Let's forget these 'charged' (programmed) labels such as psi, ESP, paranormal, and simply recognise that we are talking about higher sensory perception, which clearly should not be omitted, and should be regarded as an essentially and inevitable feature of evolving intelligence. If one person detects more of what is going on in their environment than another then obviously they could more easily solve problems, or make the best choices and embrace fringe areas of common existence, and should be regarded as more intelligent in this regard. These abilities require a functioning right brain in which the intuition is not disabled significantly.

We mentioned the ability to detect or observe quantity of data for solving a problem. This is an awareness of not only an amount of data—a span of detectable information—for solving a problem or making a decision, but the selection of relevant information. This is also an example of how complex the evaluation of forms of intelligence can be. One may have the task of handling more 3D logistic type of problems, such as in certain business or practical activities in which relevant *overt* data must be focussed upon, whereas a philosophical thinker might 'over-think' the circumstances and thus not be as successful. Thus the philosopher is manifesting less intelligence in this specific field but it may be that the philosopher has greater awareness in general, which could 'cause' too much perception to enter, in the above cases. But clearly the greater awareness represents an aspect of greater general or more foundational intelligence, in particular, in the individual's own field.

Thus where there is diminished perception of the more practical operator, there may be more benefits in specific intelligence applications. The practical person may be more positive, but is so, owing to limited data that nevertheless may be relevant to the problem/decision, whereas the individual, who is more sensitive to a wider spectrum of energies, may be more vague—too much data may be drawn in for an evaluation, requiring a more simple solution. A person spanning too much information may become confused (fusing of too much data).

The ability to think outside the context of a subject, which means outside programmed thinking, such as in dogmas, is an important intelligence in a society that is mentally manipulated—this is more than 'common sense.' It is the perception enabling one to recognise that an apparent closed system of knowledge is in fact open to a greater integration.

Can we narrow down the above list to fewer more basic categories? We can see that some abilities appear more qualitative than others, which are more quantitative. Before discussing this, let us present some relevant physics background.

Only physics can give a proper objective understanding of phenomena and enable this understanding to be communicated to another and be potentially understood (at our level). Thus agreements about what is true can be achieved. The individual may also, of course, evaluate truth by intuition but this is no good to someone else unless they also exercise a similar intuition.

Physics, when sufficiently advanced, will be recognised as the basic subject for an understanding of how things work, since we are beginning to understand that all phenomena and expressions of existence are in effect energy patterns. Thus if we understand the physics of behaviour, of nature's computer system, how everything works, we would be better able to evaluate intelligence. In fact we may find that there are degrees of intelligence which may range down to a single atom. Quantum physics indicates every particle is under continuous creation (though reveals this somewhat cautiously). We might surmise that even an atom is capable of modes of behaviour and may respond according to what operation is performed on it, or what interference is encountered. This would be intelligence but containing and based on simple programs. The simplest program and intelligence could be the sine wave (that is, what it represents—a particular oscillation, to maintain a condition).

Let us now look at the nature of quantity and quality. Quantity would be a group of separate parts but this would include parts (particles/atoms) held together by forces. The basis of this was dealt with in Section 2. Quality would be true unity which would be when the parts are in unison (in phase, resonance) forming a single whole quantum state. Quality is thus putting special order into quantity.

There may be a problem with the word 'quality' and its use here. 'Quality' tends to be used very loosely in our society for anything in which a standard is being evaluated. We are using it here, in a specific physics sense, in particular, the degree of unity, in the sense that the parts are harmoniously related. This condition would be found in music and art, also good behaviour, and moreover any entity or body in the universe, in which there is an underlying unity, such as a cell, planet, star system, galaxy, etc. (Keep in mind the undivided whole aspect of unity.) The quality of a good painting will be in the ordered relation of the bits of paint and will be a whole energy state in the mind (the aesthetic concept) when appreciation occurs, but we can hardly consider that the quality of, say, paper is an energy state. 'Quality' has very wide usage, as mentioned, but many qualities from the physics viewpoint are quantum whole states of energy in themselves—and this may apply to far more examples of quality than is likely to be believed. For example, one might consider the quality of life to be poor—this is simply a deduction—but if a sufficient number of people focussed on this state of mind it would become a thought form—a whole energy state.

A brief look at the subject of morals might tell us that behaviour is governed essentially by one's perception of other's viewpoints (ethics), and therefore this would be a mark of intelligence (it would also include one's own reactions to being at the receiving end). It wouldn't be the same, however, for a person to be trained to behave morally to a degree that there is little or no free will; in other words, is compulsive—this wouldn't carry the same level of intelligence, even though outwardly it may look the same.

Quality then, in terms of physics, will contain whole quantum states (true unity). An application might be the performance of a genius in, say, sporting activities, such as football. It is possible the genius at some point in his development does not have greater skill or athletic ability than other able sportsmen and women (within the quantitative elements comprising what skilful performance is). However, this genius will experience fleeting moments of higher unification (qualitative) of the parts, including the environmental data forming an instantaneous pattern of behaviour, which now spontaneously controls this behaviour. A synchronistic pattern of events and movements have been fused into one whole, which includes the target and success. Although we state that this genius footballer may not have higher (quantitative) skill than other best players, what generally happens is that these unifying qualitative states of higher consciousness, in producing flashes of extraordinary coordination links for complex sequences of movements, then bring about extra development of the learning patterns. These extra moments of genius come from the soul aspect rather than the 'robot' aspect of the human being.

Thus we see that qualitative and quantitative traits are quite basic in our quest for understanding intelligence. Each will clearly have fundamentally a different kind of intelligence but they combine inevitably in different degrees. An inner qualitative state, capable of achieving greater integration of the quantitative components, enables that extra special talent to manifest. The quality is the achievement of harmonious order within quantity.

Now quantitative intelligence would be more third dimensional, which means abilities controlling the relationship between parts; such as with the intellect, analysis, logic, and also physical abilities. Language would be quantitative regarding control over words but underlying the language, such as in poetry or novels, creative ability would bring in unity and the qualitative aspects. In fact each word becomes (in the mind) a whole energy state. Technique in playing musical instruments would be quantitative but musical ability is whole and qualitative. Thus abilities which can be learned might be categorised as quantitative since they are built up bit by bit, such as physical skills, or language skills. Talents would be a qualitative category—something inborn and inherent and which cannot be learned (blocks may be removed though to make it appear that it can be learned, such as in artistic abilities). Note that quantitative abilities are based on past knowledge, are learned, whereas qualitative are in present time, acting now—such as musical ability or intellectual (non-deductive) cognition. Thus 'qualitative' relates to the soul (the input to the human 'robot'). Recall that the soul is a larger whole 'portion' of the Absolute, but is still thoroughly formatted by higher aspects (fractal levels all the way back to the Source) and at this level is formatted to be a human, whereas the robot (body, brain, 3D mind) is structured from smaller portions (atoms, molecules, coherent states) and programmed to play whatever role they play, there is no large unitary beingness connected all the way up the hierarchy.

Quantitative forms of intelligence might be argued to not require consciousness; for example, a robot can be considered to manifest forms of intelligence but they do not have any whole consciousness/soul (at least at this relatively low level of robot creation and elementary simulation of human behaviour*). But qualitative forms of intelligence can more easily be seen to require consciousness. As we have repeated over and over, quality of this kind has true unity. True unity is a whole quantum state of energy and can only logically and satisfactorily be explained by relating it directly to consciousness which will be found to be a spectrum of energies (mainly scalar waves) acting in unison, with multidimensional aspects. 'Unity' in physics hasn't remotely been understood; that's one reason why it has been ignored. [* The question of when is a robot sentient was described in volume I.]

An intellectual structure is made up of parts joined, but an emotional state is wholeness—to take only a part of this energy is meaningless, except that it would give the whole in reduced intensity, and again we arrive at the conclusion that wholeness or unity is holographic. Qualitative is more experiential and emotional (but not meaning lower, or negative emotions).

Tentatively at this point we see that qualitative and quantitative appear to be basic forms. Our educational system tends only to recognise essentially quantitative values, such as logic, memory, solving analytical problems, etc. This is the ability to manipulate parts, to see the relationship between separate elements and put order into them: to apply logic, to analyse; to break down into parts for further comprehension, and also to synthesise. Basic intellect will operate on deductive abilities, but if it is to bring in new information it must combine with the qualitative.

Qualitative would be the ability to manipulate wholes; to sense meaning without reducing a thing to parts. The whole is always greater than the sum of the parts when these parts are manifest materialistically: a whole cell will be a quantum state, or a whole planet, and these will be greater, that is, contain more information than the sum of the 3D particles, molecules, etc. (due to the hidden underlying nested dimensions). In a good work of art the understanding (the ordered bits of paint) is a mind-energy state, which is more than that conveyed by the separate bits of paint.

We might immediately see that the above qualitative and quantitative factors are not independent; both are contained in almost any activity. Thus we are dealing with degrees. In quantitative or intellectual analysis, a perception of new relationships of the parts leads to a condition for a new functional state to operate—a solution to a problem, which is a separate energy pattern from the parts. We see that left-brain consciousness relates to the quantitative characteristics: intellect, logic, analysis, and objectivity, representational, indirect and non-experiential knowledge. Compared with the right-brain consciousness, which is discouraged in

New Science

our education and science today, and which relates more to intuition, subjectivity, direct, emotional/experiential knowledge and imagination. This is the cognitive process which experimental psychologists talk about—two or more parts, ideas, bits of data, information, fuse together to form a new whole—this whole is not made up of parts stuck together; if it was it wouldn't make any sense. See section on quantum regeneration.

We could say that any element which has even a simple automatic repetitive program has a degree of intelligence, or contrarily we could decide that intelligence must include the ability to form links in the program in a creative way (not automatic), which means deliberate thought has intervened in the quantitative, mechanistic process and formed a new relation. 'Deliberate thought' is a measure of the ability to take a course different from the one being presented —can a hungry animal think 'I won't eat now but later' due to, say, a weaker stimulus needing attending to first. But even this implies the perception of a bigger picture, more data to correlate, and one could again say that the bigger stimulus is in fact controlling the behaviour. Note that this forms a potential hierarchy going from a small span of attention to the greatest as one whole, indicating a single source. This brings in the subject of free will, which was pursued in volume 1, and the potential for the whole fractal hierarchy of consciousness.

The model that is being proposed here is that qualitative comes first and manifests in the form of whole quantum states from the minute (like a physical quantum of energy) to a universe or Source/God. These qualitative whole states impinge into our 3D (from inner and higher space). This is clearly a digital effect—on and off—and would have to be since the quantum states (of unity) are not of space-time (see section on fractals). This means all natural entities such as an atom or planet or galaxy, even though made up of particles as observed in 3D, have whole states underlying the particle representation (note that some quantum physicists proposed this long ago).

Art and music are expressions of the qualitative unity, but smaller quantum states—selected paint-brush strokes and musical notes—must be used to bring in these artistic concepts.

The reader may have noted that in comparing quantity and quality we have been keeping them separately defined, but as we give examples and applications we see that larger whole qualitative states, such as cognition or musical appreciation, are expressed lower dimensionally (in 3D) by smaller qualitative states (that we have been placing in the quantitative category—for simplicity).[1] Thus we could say it is all qualitative but the intelligence of existence is how large and extensive (ordered or organised) are these quantum states. However, the smaller parts are building bricks of the qualitative state (function) and are objectified, separated away from function (even disowned). Scientists will eventually establish that the external environment is in fact an objectified mind extension as indicated in quantum physics. Thus we still must draw a line between quality and quantity. Fundamental intelligence is degree of order.

Is there a more basic condition that will help us here for the evaluation of intelligence than even quantitative and qualitative? Can we reduce the 'software,' conceptual aspects of qualitative and quantitative to a still more basic hardware approach? At what point does the quantitative become the qualitative?

In addition to 'qualitative' and 'quantitative' the concepts of function and structure give us another angle on this approach to identifying and evaluating intelligence, and a better grasp of the energy relationships from the point of view of control and responsibility. Thus function and structure could be modelled as the better starting point for the evaluation of modes of intelligence. Science makes no qualitative distinction between these states (function and structure), essentially because the input to the brain or mind is not considered intrinsically different from the mechanisms of the brain or mind. One of the subjective reasons for this is that the input function, when utilising energy patterns of the mind, will fuse, become holistic with these structural patterns, forming one whole state. There is a potential feedback from the structure to the function, dictating to the function. See Figure 8.

Thus consciousness, without introspective training, cannot separate the two, and since structure has now moulded function, consciousness thinks it is structure—this is the state of science today. Thanks to mind-programming and education, the disabling of right-brain consciousness has led to the inability of the left brain to perceive that consciousness is moulded by structure—but isn't structure. In a nutshell: Input consciousness or function creates structure, and then structure feeds back information to assist consciousness. This constitutes the basic universal cybernetics mechanism.

Function, in its raw or nascent state (the Absolute), could be initially modelled as a blank slate (in terms of having no patterns of information; just an energy spectrum). There would be no rigidity, frameworks, restrictions, or any kind of limiting pattern, just complete flexibility, analogous to, say, a volume of fluid, such as water (the ocean analogy). In reality, however, this function, with its native awareness, would be given basic programs, such as a framework to preserve a separate personality (from the collective), or to survive, to have instincts, or to have certain fundamental goals; all these would be archetypes. Thus the function (as the individual control centre) now projects energy to form new patterns (of intellect or skills) based on the fundamental and broad directives of these higher archetypes which precede any learning or existence. The developing structures (such as recordings) can be thought of like imprints, grooves, or a mould into which passes the infinitely flexible input, and moulding the latter, imprinting the information within the input (we have covered similar principles in earlier sections). All learning is thus 'moulding' the input—like water being poured into rigid templates. (Scientists are going to find that information is shape of energy plus frequency.)

In the theoretical extreme, intelligence characteristics will be quite different for function as opposed to structure. Function would relate to quality and would not have space-time properties. It would (and could) only enter the space-time domain of structure as quantum states—whole undivided (small or large) energy states. Structure would be of parts, or in the extreme minuteness, points of energy strung as linked-associations, or information coordinated together. It gives rise to external and objective characteristics. We see the possibility then that all intelligence modes could be a combination of different degrees of structure and function.

Now intelligence of certain activities, both skills and intellect will increase as structure acquires more information and feeds this back to the input function. On the other hand, intelligence will also depend on the opposite condition, the degree of free will that function has. It is thus a different kind of intelligence: structure gives the quantitative, such as immediate specific information, and function gives the ability to put order into structure, to exercise imagination, cognition; to form the links. What happens if function is sufficiently subordinated to be permanently dictated to by feedback from structure? When this is from choice, such as executing skilled movements, it is possible to observe this condition (with practice). It will be found that, in fact, there was awareness of full details of the experience. This is the automatic awareness covered in Section 10.

We can deduce that free will tends to increase with creative awareness, which is also the amount of data or dimensions consciousness is spanning (this is meant at the most fundamental level of consciousness). Clearly, intelligence correlates with free will. We are not the product of predeterminism as postulated by earlier physics. But as we can see from the discussion above, free will and therefore intelligence increases with span of awareness (of information) and relates to the quality factor, function. And vice versa, intelligence is less if we have mental aberrations, dysfunctions, etc.

It is interesting to note that the materialist is not actually contradicted by this, since highly sophisticated mechanisms can still explain the above; for example, the altruistic impulse, which overrides self, could now simply be the new dominant stimulus based on a larger span of data. That 'selfish' self on limited data is a new self on the collective level of the greater data. Thus this is now the strongest stimulus. All that is happening here is that the materialistic interpretation, in approaching infinite complexity, is becoming close to its opposite viewpoint, a non-mechanistic, creative, spontaneous free will, since restrictions and boundaries are reducing towards the whole. This must be a basic form of intelligence: the degree of free will the person has, in terms of the amount of data they can accommodate at any one time with free choice (not dictated to by structure), with some ability to transcend their normal dimensional existence and tap into higher information. The difficulties in understanding free will were discussed in volume 1.

We can also see what is meant by moulding the mind. It is literal. In this case, the consciousness or function is under the automatic awareness since it is being ruled by structure and hidden contexts. This programmed condition will still manifest certain categories of intelligence but will be context-dependent (the product of our educational system is fairly typical of this). Structure is imprinting its pattern on function and controlling it even before it begins to think. In this case, the energy source of function is biased away from its ideal zero—like a scientific instrument which hasn't been zeroed. Thus this intelligence, when evaluating something new or far-fetched, will be useless—it merely makes a comparison between the new information and the existing mental contexts.

We can't overstress the fact that man has been programmed to completely underestimate his potential and the complexity and intelligence of his dimensions of existence (internally). Physics could release this limitation with a proper recognition of quantum physics, chaos theory and fractals. These will lead to the establishment that human consciousness is an entity in itself and has higher aspects (fractals), all of which interact holographically (but with dimensional and programmed restrictions), and these higher properties must at least, to some degree, be included in intelligence evaluation.

Some scientists, and other sources, are developing quantum physics towards a unified field of one type of basic particle, called a consciousness unit, initially. If all particles are under continuous creation and are clearly regulated by programs and/or codes, then we need a new definition for life. The meaning 'organic' will then be a sub-category. Thus on this basis everything is alive and has intelligence and we should respect this. Children should be taught *reverence* for all life. However, this can be defaulted to only our recognised life forms (all creatures), since including the rest (plants and ultimately all manifestation; every atom) would involve a much more advanced viewpoint on creation.

Furthermore, another stumbling block to the problem of evaluating modes of intelligence is that science and the development of knowledge on this planet, generally, does not yet recognise past lives or higher-sensory perception. We have no choice but to ignore these suppressive restrictions if we are to honestly evaluate and theorise as to sources of intelligence. Are we so naive as to think a 'Mozart' began his life on Earth with a clean/blank slate (just genetic inheritance)? Moreover, a number of scientists have established positive evidence (but suppressed) for the tenability of past lives and therefore continuance of life[2] (which has been proved countless times in other fields). Special abilities such as these cannot be attributed to existing (current life) structures or learning from scratch in this life, or genetically acquired attributes.

In terms of function and structure it means that structure would contain programs or instinct imprints, and genetic information. Function would be the questionable item as to whether it (function) begins with a blank state. With our above argument

New Science

(Mozart example), we say that function brings in already acquired characteristics into this life. Function is the input consciousness, which has energy patterns within its energy spectrum, of which the latter consists of frequencies of higher rates (than the material level) and that they are essentially scalar waves (not presently detectable by scientific instruments). This has been covered earlier in the book.

If some intelligent forms are contextual, that is, acquired by interacting with the environment with external data that is sorted and associated, then some internal medium is being moulded or imprinted. Thus function or consciousness instantly takes on (is formatted by) the impressions/patterns of its structures which may have been gradually built up. We could say that the source of function is the awareness unit which can be regarded as a structureless medium in itself but that it will be bringing into its life the archetypes (explained earlier), plus impressions from prior (to this life) learning experience (and also higher orders of mind recordings inherent within its higher fractal levels).

A being free from compulsive structural impediments, neurosis, etc., with an ability to understand itself, its personal intelligence, would have an awareness of being aware, recognising he or she possess higher aspects of themselves (in a fractal system, that is, a spirituality, soul, etc.). This would be the primary and dominant form of intelligence—this would include higher-sensory perception (a perception of higher frequencies). It is an extension of the normal (or subnormal) range of detectable frequencies by humans. Inexcusable denial of this higher sensory intelligence occurs because of several reasons: 1) religion associates it with the devil and black magic, 2) education creates a fear and insecurity of the unknown which deflates the ego of the arrogant intellect, 3) a failure of science to recognise it, owing to a refusal to understand the nature of true unity and resonance—we may find that all learning (by the creative awareness) is resonance, bringing integration. However, almost amusingly if it wasn't so ridiculous, all the above are manipulations to conform to an agenda of restricted knowledge and understanding of evolution and the universe. This non-recognition of higher sensory perception is a deplorable omission but nevertheless is a key factor in arresting the development of a civilisation. Higher sensory perception and intuition need to be recognised to be part of the evolutionary and ascension scale of humanity's progress.

Regarding the state of affairs on this planet, one could say that function (and quality) is deteriorating into structure (quantity). One might see that this is like saying the information from the nonlinear, internal quality or function is focussing on less, and the external interaction of separated parts is more dominant. More and more we are evaluating life on a quantitative basis—this is a certain deviation from an ideal and what should be an ever-ascending evolution. It will result in certain demise of the civilisation. Function is creating so much structure and relegating responsibility to structure, the latter begins to 'mould' the function and take control. Only the unity aspect, along with function, has any true existence. However, inherent within function some of those finer 'structures' which were brought in at birth, will be actual structures in the fractal level 'above', which we identify as the soul level—and so on to higher fractal strata. This will go back in fractal levels to the final condition of a theoretical blank slate of no structure (the Absolute). This would mean that the theoretical native state is entirely qualitative and operates on whole quantum states. It is singular and thus has no quantity or 'external' properties. However, it brings in nonlinearly the internal dimensions—the within, the inner self.

Note that we have been viewing 'quantitative' mainly in a linear sense. If we extend the 'quantitative' into nonlinearity and inner space, that is, internally, this quantitative condition will simulate the 'qualitative'. If we imagine a company organisation again and consider firstly the ground-floor workers in isolation, then we are demonstrating quantity. But if we include the managers, executives, and president hidden behind the ground-floor workers we have introduced internal nonlinearity and have simulated quality more, but nevertheless, the unity of the true qualitative state is necessary to link together, unify and organise the 'individuals' (into whole states of energy provided by function). With our present computers these internal levels (managers, executives, etc.) would fuse together and not be differentiated from the functional and qualitative point of view, that is, they couldn't be handled on the same level (ground-floor workers)—without frequency and dimensional separation.

So where do we draw the line between quality and quantity? It depends on how 'local' is the region chosen; in other words, it depends on limitations. The qualitative and functional aspect puts the order into the parts. If one isolates a biological cell one could say that the quantum state (unity) of the whole cell is acting to some degree in a qualitative manner over its particles. But in the context of the whole organ, a higher level of control, the cell relegates its control to this higher structure (note the dependency on context). Man, in his restricted 3D, is function over his 3D life and environment but some of this qualitative state is overridden by his next higher fractal level of consciousness. Thus again we see the contextual gradient is fundamental to existence—this also clarifies the subjective/objective relationship puzzle.

In effect, a single whole (source, God) has least limitations and maximum 'quality', but as we increase the fragmentation of the One—that is, forming a fractal system—limitations increase to the degree there are separate parts (one part doesn't know another part—except nonlinearly through the whole (via inner/higher space)). The general principle is that the smaller parts follow the patterns dictated by the larger parts on a contextual gradient back to the whole. But the smaller parts can have freedom within the fractal restrictions.

From the ultimate intelligence point of view one *knows* to the degree of (being) the One whole or to the degree the parts are brought together and integrated into, ultimately, wholeness (but the whole aspects are already there—higher in the dimensional scale—and are quantum regenerated).

There is another aspect of intelligence applicable to more advanced civilisations but nevertheless a feature more common than one might suppose in our society. Because we emphasise objectivity, physical senses, material reality, so much on this planet, we do not recognise two paths to achieve a goal—two types of intelligence? That extra talent or genius is the qualitative state of integration/connectivity, bringing about the perception of end results and allowing parts to align to achieve goals. The target or end result could be almost anything: geographic, any achievement, a skilful movement, events which lead to a desired conclusion. Whether we know it or not, two variables, two factors interacting are involved in most tasks. These are 1) operating via the parts (structural and quantitative), and 2) operating via the whole, or single quantum states (qualitative and is a functional aspect).

A geographic example would be the intention to get from, say, X to Y. The normal route would be by breaking down the whole into parts—looking at the steps leading to the target and applying deliberate action to handle these steps, in other words, using a map. How could it be done any other way? It could, and possibly animals use this ability. We all know of the animals that find their way to a location without knowledge of the route. If the animal is, say, 100 miles from home and mentally envisaged being at home, a frequency pattern is activated corresponding to the home location. Any two or more resonating frequencies pull together (frequencies of the mind—mainly scalar). Thus the pattern in the mind of the animal now resonates with the actual locality of its home. It will feel a pull in that direction.

We see then that achieving a goal on this basis is governed by the target focus or end result, causing the intermediate parts to align automatically. A skilful movement operates precisely this way. There is never attention and individual control of the parts in high skills (the control is nonlinear). In general, man mainly focuses on action along the path to the goal (works out the route; breaks down the whole into parts) except in activities such as skills in which the whole idea of a movement is formatted into computer bit-patterns to achieve the result.

Basic or universal intelligence of creation itself is demonstrated by the fractal and ordered nature of the universe. A fractal system goes from low order (e.g., twigs) to higher order (branches, trunk). All other forms of intelligence, such as the IQ test, are subsets of the universe's ordered structure of dimensions for the evolution of species. The physics of this system of orders (organisation within organisation) means all observing entities (scientific instruments, life, or even any energy) experience an environment based on their reality or context. If there is an interaction with a higher order, the observing wave patterns will quantum-reduce the greater order or coherent state to their own level. Science unwittingly, in particular, in experimental methodology, collapses the wave function of the higher order to the lower level—similar to the idea of analysing unity (a higher truth) and breaking it into parts, reducing it to a lower level of truth (Section 3).

We have looked at how the mind works in the application of intelligence, but how does the universe work? Meaning here, how does it present itself. Does it use, for instance, abstraction, such as algebra: Let x equal this or that? This is similar to our computer binary system with the abstractions '1' and '0', representing bits of information; a contrivance (clever) of the intellect. What if a religious scientist says, Why shouldn't a Supreme Being choose an intellectual abstraction? The simple answer is that this is like algebra in which a preference (Let x equal . . .) is given to one of many options, equally as good. This is a failure in the tests of truth of physics (Section 4), which states that preferential formats should be avoided. However, there is no problem with geometry and so we very much favour (even without the practical evidence), that the universe functions on geometric intelligence. It is all hardware (patterns of energy and frequencies).

We have previously covered the subject of geometric intelligence in volume I, here we shall merely present a recap and summary. Note that we have frequently related function to the qualitative, and structure to the quantitative. But if quantitative relates to geometry, which it will, we find that the universe works on geometric intelligence (frequency/wave patterns) and we also have the potential for the qualitative with geometry. Science recognises that geometry is an absolute system, and therefore we can have both quantitative and qualitative with geometric intelligence.

Thus we have two more classes of intelligence (but which overlap with those discussed): geometric and algebraic. Note that the left brain (thinking, intellectual, theoretical) is associated more with algebra, and the right brain (looking, practical, intuitive), geometry.

As an additional feature to presenting evidence for the geometric approach we can include the ocean analogy, given earlier. The ocean represents the absolute, infinite, stillness or Source, and the wave/vortex patterns on the ocean are the manifestation of all quantifiable phenomena, matter, thought, emotions, mind and consciousness, where these are expressed in particles and waves. In this analogy we can now 'see' the modulations of the ocean forming the wave-pattern structures, the basis of an existence (but we can't see, and science can't detect the 'ocean'. The modulations are shapes, something moulded, and we have information and intelligence from geometry.

We see geometry in a fractal system of universes (see earlier sections) with ordered dimensional boundaries regulated by universe levels (3D, etc.) which are relative to one another, which means each has its own relative zero (reference), decreasing in degrees of relativeness up the hierarchy to finally the Absolute zero.

Infinity is exploring its infinite possibilities by becoming finite. There is a fractal division in the cosmos of universe dimensions of decreasing degrees of order as we move down the scale. Lower order life forms cannot ascend to higher levels unless possessing the necessary qualification. Lower order knowledge cannot detect higher order knowledge. The latter will quantum-reduce to the lower level (collapse of the wave function). Similarly life forms due to either, or both, insufficient frequency value or consciousness expansion (wholeness, integration/holiness) which must be manifesting on the lower level before ascension, will not qualify. Besides keeping order, this protects the integrity of the cosmic structure by preventing a lower degree of order/coherence/harmony from mixing with a higher degree and bringing down the higher order, causing eventual destruction of the fractal system.

Our intellect and computers utilise abstractions, such as algebra. As indicated above, in computers we have the binary system with information bits '1' and '0'. With geometric intelligence it is all hardware and there are no zeros except the Absolute 'stillness'. There is no preferential control; geometry is an absolute. The wave pattern has frequency (rate of information) and geometry/shape/pattern. Certain patterns of pulses in space and time form codes, and operational information generally.

All geometries radiate wave patterns and our subconscious can read them when these frequency patterns are in correspondence (that is, when using a language of the universe or creation, such as in DNA). Communication and 'understanding' occurs by resonance.

Thus the information in our DNA/cells is part of our subconscious and can read incoming compatible patterns of information. Our 3D conscious mind and intellect can't do this. Thus anyone advanced enough using bio-genetic physics technologies could send information to our DNA. This is how crop circles work (human creations radiate garbled information).[3]

Notes
1. This is like saying objects (even a robot) are made up of tiny parts (say, atoms or bio-cells) and do not have a true whole (quality)

intelligence, but in fact each small unit, an atom, cell, has its own wholeness or quantum state and therefore has a tiny piece of the qualitative or intelligence. Also don't forget quantum regeneration here; this will generate, by resonance of the parts, a natural wholeness, but more like a collective, for example, cells of the body and the whole body consciousness. But this is not the additional soul personality.

2. Book: *The Emerging New Science*, vol. I, past lives.

3. Books: *Voyagers* vol. I, and *Engaging the Extraterrestrials: Forbidden History of ET Events, Programmes and Agendas*.

PART FOUR

SUPERSPACE

19.

THE UNIVERSAL VORTEX AND ELECTROMAGNETISM

The universe is held together not by gravity but a network of vortices, which provides a perpetual energy circulation system, and also generates matter and all bodies, at the centre of pairs of dual counter-rotating vortices.

Modern physics theories now recognise space or the so-called vacuum as a very active medium saturated with quantum energy fluctuations, which has been described as a turbulence of particles, mini-black and mini-whiteholes, and a sea of 'bubbling' foam. Even prior to this recognition of space as not being empty, the aether was postulated into existence to provide a ubiquitous medium through which waves could propagate.

Keep in mind the ocean analogy as we describe the characteristics of the structure of space and the aether; the water waves and vortices of the ocean represent electromagnetic waves or particles. Similarly the aether will have a fluidic flexibility. Also many sources of information indicate that this aether has elastic properties, that is, it can be stretched and compressed, giving rise to tensions with the subsequent tendency to equalise these differentials by the flow of the medium (aether) from higher potential to lower. As a finer point these flows and motions will result from the passing on of disturbances; similar to a wave, which doesn't actually move linearly but passes on the undulations. This turbulence of rapid and reinforced motions will cause the well-known mini-black and whiteholes; these are considered virtual particles and not observable mass, such as an electron.

It might be of some assistance to initially present a vector model for this turbulent aether. In Figure 20, we have illustrated a random grouping of vectors; they each have direction and a

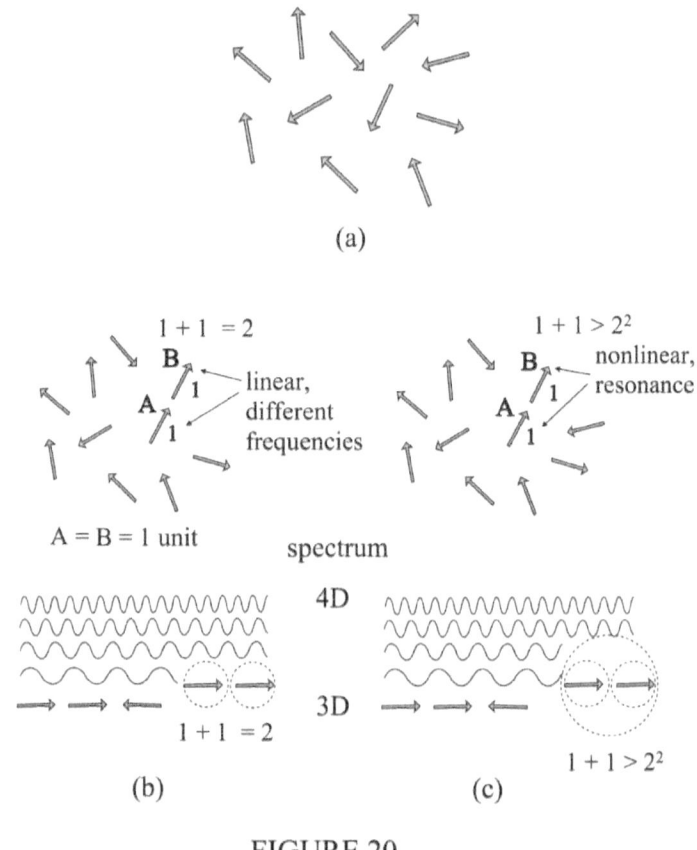

FIGURE 20

magnitude, such as a certain value of force (usually represented by length). These arrows represent for us the basic turbulence in space-time, that is, the most fundamental level of energy interactions, which has been described as the virtual state; also referred to by professor John Wheeler as a quantum foam of bubbles and wormholes of geometrodynamics, and subsequently a sea of mini-black and mini-whiteholes by Bob Toben and Jack Sarfatti.[1]

We are proposing that all things emerge, or are constructed, engineered and programmed from this framework (the aether). Keep in mind quantum physics conclusions, that all bodies and objects (all manifestation) appear to be the result of variations and shapes in the structure of space. Recall that quantum physics was on the right track here but relativity (with a lower degree of order, as per the tests of truth) brought about the removal of the aether theory, causing quantum physics to now contradict itself regarding the definitions of mass, energy and force.

Now what are these arrows in the diagram? They represent particles generally and mini-white and blackholes of fleeting existence, moving in all directions, covering a range of frequencies. They are all essentially nodes and appear like oscillating particles. During the continuous motion of these vectors there will be moments when two or more arrows line up, say, in close proximity, for instance A and B in Figure 20(b). There will simply be a resultant vector which is equal to the sum of A and B. This is a regular linear system that classical mechanics can handle. But now what happens if, in addition to alignment in direction, the frequencies of A and B are coherent? (Figure 20(c).

The effect is quite dramatic—it is commensurate with any of the startling features of chaotic systems. What happens is that owing to the presence of the virtual hierarchy or spectrum, to be explained shortly, which is behind all processes and extends into higher frequencies and dimensions, in-phase energies on any lower level regenerate a more coherent and integrated higher-harmonic frequency (compare the laser in Appendix B).

This is a process that is consistent with the overall theory given in this book and, in particular, the phenomenon of quantum regeneration is also a logical accompaniment but reciprocal to quantum reduction. This resonating region is a greater whole, a gestalt of the vectors on the lower level. If only two vectors are coherent then a level only slightly higher will be regenerated. The two levels form a polarity relation and are in harmonic resonance but the upper level has a higher frequency (use the triad principle to see the relation between the two levels). Note that the quantum wave function already implies such holistic levels. If more vectors are involved then a greater whole oscillation is kindled of greater dimensions along the gradient into higher-dimensional spectra.

This continuous process becomes immediately enormously complex even in the simplest system—resonance and mode locking coming and going at different levels of unification. In Figure 20(b) and(c) each vector A or B is given the magnitude 1 unit (of energy). We see that the new vector in Figure 20(b) is simply A + B, that is, 2 units. But note that we should also put in the thickness in the 4th dimension (a quantum, for example, is not just spatial but has an equivalent thickness along a 4D direction, that is, along time). Now in Figure 20(c), since A and B are also in phase, frequency-wise, an extra 4D thickness is created owing to the greater diameter of A + B; this is quantum regeneration explained earlier. We can now replace all these vectors by mini-black and whiteholes or particles generally, which would exhibit a similar behaviour.

Now where does all this superspace of turbulent activity come from? In the quantum 'vacuum' every point is actually an infinite source of energy, though quantum physics limits it to a finite, though enormous amount, since current quantum physics does not recognise the inner-space fractal hierarchy all the way to the One Absolute, which is eternal; a perpetual source.

Keep in mind again that quantum physics reveals all things appear like shapes and motions in the structure of space, as described in the ocean model above. The flowing and spinning aether will stretch and compress. An oscillatory or expanding region will be less dense and a region of more activity, such as where there is wave reinforcement, and where there is cancellation of waves there will be stillness. In these latter regions, 'condensation' of aether can occur. Nevertheless this is a 3D description and more precisely the waves are around the nodes, of which the latter are the mini-black and whiteholes with a resultant flow direction at right angles to our 3D, but spinning (that is, they require 4D to describe them). See later figures.

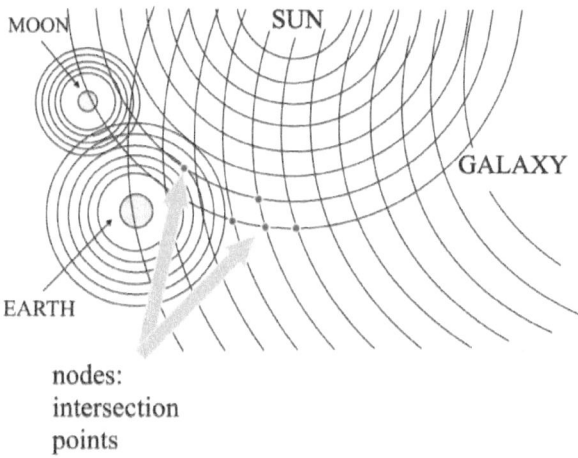

nodes:
intersection
points

FIGURE 21: Creation of space particles/-nodes from countless intersections of waves from celestial bodies.

Recall the fractal hierarchy described in earlier sections of dimensional layers and countless celestial bodies and galaxies, etc. All these wave propagations create infinite intersections and interference patterns over a huge range of dimensions (since as we have explained, larger natural bodies have higher orders). All this wave activity from different dimensional levels of the aether, giving differential pressure effects, act in a manner to cancel and reinforce wave crests and troughs, creating still zones or zones of greater activity from these wave reinforcements. There is sufficient violence to puncture the dimensional medium under consideration, giving not only the mini-black and whiteholes but where there is a region of many intersections involving cancellations and stillness, we have the appearance of mass particles due to contraction and condensation of the aether. This activity of mini-black and whiteholes is the effect of the underlying countless potential gradients of aether densities/tensions superimposed from these waves/oscillations of various bodies, such as planets, stars, galaxies, and universes, extending into the multiverse hierarchy. See Figure 21.

As a result of all this interference within the aether, countless nodes are formed (standing wave structures) from the intersections of waves/'ripples' from the complete range of celestial bodies and natural structures. This is a complete hierarchy of space-time dimensions of increasing spectra of frequencies ranging from small particles to planets, stars, galaxies, etc. Note that higher-dimensional effects will project into the lower levels, similar to the analogy of the spotlight phenomenon (Appendix D).

As indicated, matter forms where many nodes coincide and is thus merely a condensed configuration of patterned aetheric motions. We could say that the aether consists of two states, 1) an inactive absolute medium of infinite possibilities and potential (see section on Absolute) and 2) this same state disturbed (or 'moulded') by a myriad wave patterns from celestial bodies or significant natural masses in general. Thus these cause the countless nodes and fleeting mini-black and whiteholes or what are called virtual particles, but also observable particles; recall the electron 'whorl'.

The basic source from all these waves would be from the principal vortices forming all celestial bodies; even a universe would have an enormously high frequency everywhere within its space. Although there is not yet any orthodox acceptance today, there is an overwhelming amount of data supporting the vortex theory as fundamental, from not only peripheral scientific research, exploring the frontiers of orthodox science, but investigators within the fields of religious philosophies, New-Age transmissions, and metaphysics. The scientific community is programmed to think there is nothing worthwhile in these fringe practices and simply haven't realised this knowledge is being suppressed.

Investigations into the understanding and application of the vortex has revealed unexpected and quite startling physics when compared with orthodox science; such as the creation of machines exhibiting anti-gravitational effects and also excessive energy outputs above that required for the input. The vortex has a spinning, spiralling motion, with similar fluidic properties to the whirlpool or tornado. It has electromagnetic characteristics created by its type of motion and shape within the structure of space, which we are addressing as the aether. Recall again the ocean analogy of early physics, in which a tiny whorl in the surface of the water has the character of an electron. As stated, the electron is not a thing or particle but a hole in space and draws in energy, which is then spiralled out of the 'surface' into 4D (hyperspace of higher frequency), giving the property of the negative charge.

We say that the mass of the particle, body, etc., forms at the intersection of the vortex pair, but what do we mean by 'mass'. See Figure 22 (basic vortex).

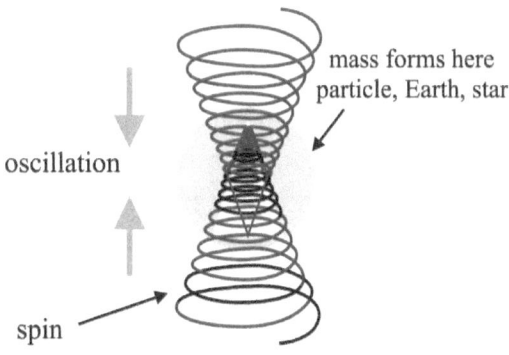

FIGURE 22: 3D dual-vortex polarity.
But also exists in 4D format.

The understanding of mass in science is quite contradictory. In the New Science everything is formed out of patterns in the aether: waves, whirls, eddies, etc. The mass is merely a structural pattern in the aether. If the mass is freely moving with the aether motions, whether a mass is a particle or planet (in orbit), the mass is coherent with its environment. It does not manifest Newtonian properties. As described in some detail in volume 1, when measurements are made of the body in this free-fall state, a (secondary) collapse of the wave function occurs, which knocks out its coherent state with space, *then* manifests properties of inertia, kinetic energy, momentum, and therefore mass, which is thus a special case (see section on gravity).

Quantum physics recognises some coherent states of matter and energy, such as in the Bose-Einstein condensate and, in particular, in the field of the development of quantum computers in which the coherent state must be preserved (that is, to avoid the collapse of the wave packet) to create the quantum computer.[2] The Schrodinger wave equation only describes the end result, that is, after the collapse of the wave function, when the selected probability has occurred. As quantum physics expands into the inner-space fractals of greater coherence, it will require a more basic equation without the Newtonian mass term (that is, as mass is defined today). This more primary equation would represent the coherent state (the result of integration of the Schrodinger equation) prior to the collapse. It is a harmonic state in which the quantity representing mass is in a phase-correlated condition. It does not have the properties of force, inertia, kinetic energy, etc., and therefore mass. The equation would then be differentiated to give the phase-randomisation condition and the current Schrodinger equation (the result after the collapse of the wave function).

Theoretically, increasing coherent states should continue up the fractal scale, up to the quantum realm (and basic 'collapse' recognised today, expressed in the Copenhagen Interpretation); in other words there is a gradient here of quantum reductions within

The New Science

quantum reductions (see Figure 38, Appendix D). Science only recognises one level here. However, it is doubtful that mathematics could ever handle this. Let us now take a look at the vortex as fundamental.

For stability, the vortex functions in pairs with counter-rotating motions. The dual vortex configuration is the universal and basic system of circulating energy throughout all manifestation, flowing across dimensions, creating portals under special conditions but, in particular, provides the energy and necessary counter-energy for the formatting of bodies, particles, planets, etc. (see Figure 22). A network of vortices throughout the cosmos provides a perpetual energy system, which has been simulated by the technologies of inventors, such as John Searl and Joseph Newmann, using electromagnetic fields, and Viktor Schauberger and Frank Polifka, using water as the vortical medium to create spin. In all such examples, with sufficient spin rate the systems generated excessive power, creating more energy output than required at the input, referred to as 'over-unity'. All such inventions have been suppressed by the authorities.

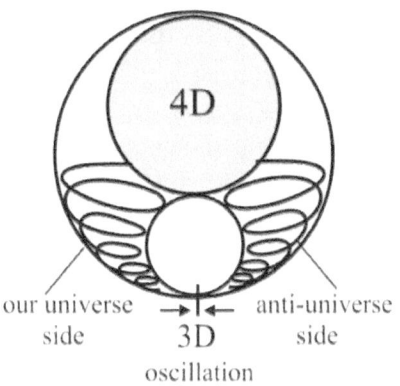

FIGURE 23: For clarity, 3D is projected outside 4D (should be inside).

Figure 22 is the standard vortex model with the two counter-rotating vortices. Note that this is a 3D depiction where both poles are in the 3D plane, such as are the poles of a magnet, or an electron and proton pair. Also the planet Earth would have such a dual-vortex system, lined up with and partly creating the North and South poles (apart from the known magnetic field contribution). Nevertheless, the circulation of spinning energy (disturbance of the aether space) will have to complete by going round through 4D—Figure 23. (This is another view of Figure 22 and reveals the outer vortex connection joining up in 4D.)

Science does not detect the Earth's high frequency of the vortex pair (much higher than an atomic frequency). Now, the Earth will also have a 4D vortex system, much more difficult to visualise than in Figure 22. This vortex pair would be functioning 90 degrees to 3D space. Only one vortex in the pair, appearing spherical, would be detected in our 3D, the other is in the anti-universe side (along a 4D direction). To visualise this 4D vortex see Appendix C. An understanding of the geometry of this energy unit in its higher-dimensional state will tax the visualisation processes to the limit. In fact it cannot be completely visualised with 3D perceptions and many diagrams are necessary, presented from different angles of lower dimensions.

Figure 23 gives quite a good depiction of the two spirals emanating out of the 4D sphere, and where the two vortices meet there is a 3D oscillation. It is exactly the same vortex unit as described but for clarity the two spirals have been extended out from the 4D sphere—they should be inside. That is, the 3D intersection point and the two vortices should be encompassed by the 4D 'circle'.[3]

The vortex system at particle dimensions would be quite flexible in its form, but for natural and primary bodies, such as Earth, its spin rate would be inherently constant, unless unnaturally disrupted. For example, antiparticles can be detected within our side of the universe/anti-universe vortex (though this may be a typical malfunction within this degenerating black-hole galaxy).[4] Thus they appear separate from their particle-counterpart but they would be connected in inner 4D space. The relationship of particle to antiparticle is at right angles to our 3D space, as is the case with larger bodies including the gravity-antigravity field of Earth and all natural bodies.

Particle and antiparticle also have a central zone, just as middle or Inner Earth is central to Earth and its anti-Earth.[5] Thus particle and antiparticle have a greater unity as they come together as the central particle.[6]

The vortex pair formats bodies at the intersection of the individual vortices; see Figure 22. It is also the source of gravitation as well as simultaneously formatting the mass of any celestial body and its motion. Thus all physical manifestation, such as stars, atoms, particles, forms at the centre of interactions of dual counter-rotating vortices. Although we won't pursue this feature, however, imagine in Figure 22 the top and bottom vortex merging together, the top one moving down and the bottom one moving up, creating greater unity between them. But preferably use the 4D model and visualise as shown in Appendix C. This is the particle and anti-particle side coming together—like two opposites forming a greater whole. This is evolution of structure—even including universes.

We could say that it is one of the most remarkable failings of orthodox science to have 'successfully avoided' the vortex model; the fundamental energy unit of existence. It was even very popular in Lord Kelvin's day, along with the recognition of the aether medium through which electromagnetic waves could travel. However, modern physics, the Michelson-Morley experiment and, in particular Einstein and relativity put an end to these vital ideas—the vortex and the aether. One might suggest that vortex research involved too advanced physics for that period of scientific development, and thus it wouldn't fit comfortably into the current limited paradigm and so vortices were abandoned after about ten years of study.

Science considers that these vortex forces that underlie tornadoes and whirlpools can be understood by current classical laws of physics: by hot/cold vacuum effects, suction power and force of air pressure, and Newton's laws. It will be found that underlying these effects is a much more advanced physics—which could bring abundant free energy. The vortex is so 'willing' to spin with enormous power once the correct conditions are achieved.

The great Austrian inventor Viktor Schauberger, in the early 1900s, was the first pioneer to understand the vortex as a natural product of nature's creations and not as a theoretical construct, and recognise its potential. Typically his successful applications were suppressed. Reports reveal that a business syndicate bought the rights with a promise to market them, but unscrupulously shelved them. His knowledge of the vortex and how to create and harness it enabled him to develop antigravity systems. What we conventionally recognise as a powerful suction towards the centre of the vortex, also will be found to curve space and manipulate gravitons. Schauberger was caught by the Nazis during WWII and forced to work for them or be executed. Towards the end of the war the Germans were apparently utilising this vortex principle in 'saucer' type craft. After the war anything of this kind of military value was confiscated by the allies—and Russia and the United States shortly had secret antigravity crafts.

Vortex theories are much more numerous now, mainly originating with the new science pioneers (usually frowned on by mainstream academics). However, all such limited vortex theories, which have included toroid-shaped vortex rings (2D-3D) and attempts at spherical (3D) vortices, have failed to interest anyone except the few and, in particular, they fail the tests of truth of modern physics, such as supersymmetry, non-preferential axes, and generalisation, partly because they are only 3D (see Section 4). Fortunately, and quite magically, as already mentioned, if one adds another dimension, the perfect geometry comes into view. It now has supersymmetry, and to our 3D perceptions it appears as a sphere with an antiparticle or anti-universe side to it. See Appendix C for visualising the 4D vortex.

As a further technique for visualising 4D, it might be a good mental exercise to imagine again moving one's consciousness through the surface of an atom and following a radius to the centre, what will happen? In 3D this is as far as it goes but in 4D the consciousness will go through the centre, curving through the hole in a 4D direction. It will appear as though we have flipped over and are returning on the same side. Not only is it the other, mirror-image side but also if one follows the surface it will return to the same starting point on the outer edge (which is in 4D), representing the unity of both poles (the centre of the atom, a region open to 4D).

Vortex theories, as mentioned, were seriously considered by some prominent mainstream scientists more than a hundred years ago. Where they failed, and were subsequently rejected, was in the lack of polarity relationship of the pair of vortices and, in particular, that the 3D view of this two-polarity vortex energy unit was inadequate to provide a basis for all phenomena. If physics took this dual-vortex system to the next dimension, they would find the perfect supersymmetrical configuration that is much sought, which reveals the anti-side of all universal structures: particle/antiparticle, gravity/antigravity, universe/anti-universe and, yes, there will also be planets and antiparticle planets, etc.

The vortex system is fundamental to all entities and is present all the way up the fractal hierarchy, from particles to atoms, cells, organs, mind, the planet, solar system, galaxy, universe, etc. These are just a few of the stable fractal levels in the universe. This vortex energy unit is basically the same for all bodies except that its parameters are more evolved in the range particle to universe. As briefly mentioned, the vortex pair evolves by uniting the two separate vortices in the centre (see Note 6).

Any vortex in the hierarchy of atoms, planets, stars, galaxies, etc., will spiral within itself (each higher state encompasses the lower) to the lower sub-fractals below—see Figure 4. A fundamental planetary oscillation will contain all sub-frequencies, such as of atoms; it is a higher organisation. It is a fractal level higher than an atom.

The vortex principle also applies not only to atoms but subatomic particles. The spiralling vortex induces a secondary vortex of magnetic lines of force that shape the toroid form (Appendix C). This secondary vortex manifests in the gradient more towards 3D and at right angles to the main vortex and displaced in time by 90 degrees. With this theory the lines of force accompanying this complex oscillation, surrounding the centre of the atom, would intersect in a consistent rotating pattern; the points of intersection creating the minute ingoing vortices of the electrons. This interaction of the vortex and its secondary will give an interference pattern. An electron will thus be found to be an ingoing black-hole vortex, formed similar to water surface undulations, caused by interference of waves. Since this is in the dimensional range 4D to 3D these interference fringes would appear similar to alternate light and dark shells in an atom (4D geometry). For the nucleus, protons (positive) would form at intersection points in regions of low activity due to wave cancellation where compaction occurs, giving rise to a white-hole effect of aether flow, coming out and flowing to the nearest electron. Both the electron and proton complete the circulation through the underlying 4D (Figure 26). Consequently the opposite charges will attempt to pull together by the aether flows (which will in turn carry the countless mini-black and white holes).

Thus the electron must be balanced by an equivalent but opposite flow, which would be the proton. These 3D and 4D flows are what we call energy, and are basic to energy and work done by particles (science only recognises work done by charged particles with no aether flow that carries the particles). Thus whereas the electron is like a black hole, energy going in, the proton is like a white hole, energy coming out. By the principle of like attracting like, the electron and proton would attract mini-black and white-holes respectively, giving a virtual flux. But as we have explained, particles such as the electron and proton will be nodes formed from the more energetic and reinforced intersection points and that the electron and proton must connect up again beyond 3D in the underlying 4D space, forming a completed circuit. See Figures 24, 25, 26.

In Figure 24 we see the effect of the magnetic force on the motion of the electron (the electric force carrier). Note the arrows when not in opposition draw the electron in that direction. When in opposition, increased compression of aether will push them apart.

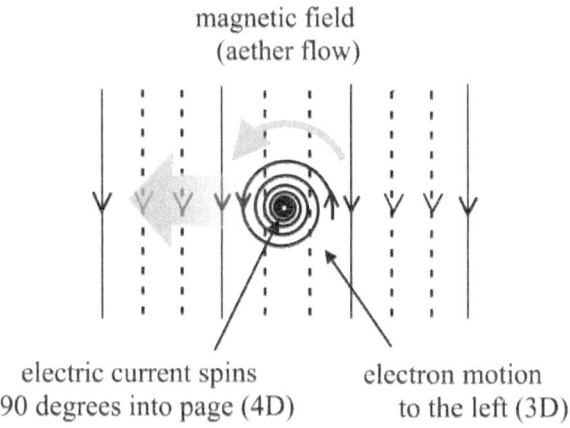

FIGURE 24: The field arrows represent direction of flows.

In Figure 25 we see the magnetic fields of the two electrons merging. In a wire carrying a current, electrons spin in the same direction and their magnetic fields will form one big loop around the wire. Note that the magnetic field's adjacent directions around the electrons are in opposition. This does not contradict the case in Figure 24, in which the magnetic fields *only* are interacting; a 3D effect. In Figure 25 the direction of the electrons spiralling inwards (4D-wise) is dominant, and not the two opposing magnetic fields.

Although we have stated that the electron is like nothing, that is, a hole, it would be expected to have finer structure underlying 3D, and a logical conclusion for the vortex nature of particles and antiparticles, say, an electron and its antiparticle, the positron, would be that the electron has the 3D vortex (giving some 3D stability) plus half the 4D vortex, that is, a spiral just going inwards (black-hole type). The positron would have the same 3D vortex but half the other pole of the 4D vortex, that is, an outgoing flow (coming into our universe—a white hole). This is speculation in an attempt to make the electron more stable.

Now from the particle and antiparticle configuration one may see why such basic particles as an electron must be 'turned' twice 360 degrees to return to their original orientation, which of course is only 360 degrees with a 3D sphere, such as a ball. This is also the reason for Dirac's discovery of negative electrons and antiparticles. (A negative electron is a negative, negatively charged particle, which is called a positron (has a positive charge) and is an anti-electron.)

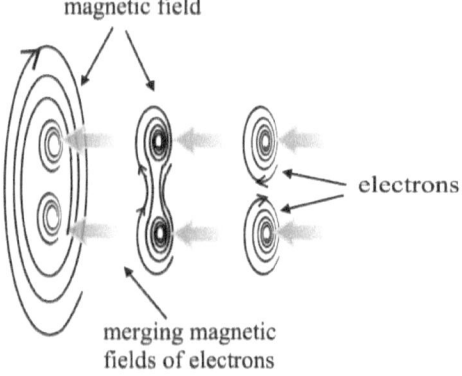

FIGURE 25: The field arrows represent direction of flow of aether. The two electrons could be moving in a wire and the magnetic field expands around the wire.

The positron can still be detected on 'our' side. The 'coin' has two sides; one side can't exist without the other. Similarly matter and antimatter, half a 4D oscillation/wavelength each, cannot exist without these two sides. We in 3D only perceive one side of the 'coin'. Also an atom or any such basic entity cannot have symmetry, for example, be spherical and be stable unless its formative energies utilise the fourth dimension. The opposite anti-side is necessary for stability and the flow of energy/aether through 4D. This basic energy unit, the 4D vortex, also explains the strange mirror-image properties found in particle physics studies.

Note that there is really no mass as such for all bodies; the electron going out is more like an absence of matter—a stretching of the aether medium due to high intersecting activity of waves, and the incoming positively charged proton (the characteristic of the centre of an atom) will manifest from a condensation effect of the aether; a pressure arising due to the low activity, that is, cancellation of intersecting waves.

In Figure 26 we show the drift of electrons in a circuit, and below we emphasise the potential pressure from left to right, and see the creation of the electric field. The flow of aether is from negative low pressure electrons spiralling inwards into 4D space and round towards the positive end (higher pressure). Note that science recognises that not only electrons (negative charges) form a current (in metals), but also protons (positive charges) provide current in many other mediums. Some materials have both positive and negative charges flowing in opposition.

Now contrary to current science, charged particles do not *cause* electric fields but field tensions create charges. Similarly mass does not cause gravity; the gravitational field causes the properties of mass. Mass is weightless in free fall. The gravitational field is not a force-field (but is scalar); however, it creates force on a mass when the mass is restricted in the gravitational field, whence all the Newtonian properties come into existence. Force is an effect, not a cause.

Science tells us that a vibrating electron is creating the continuous electromagnetic energy produced, not recognising there is a continuous input to the electron's charge of virtual energy and mini-black and whiteholes.

Electrons are moved by aetheric potential differences. An electric charge (static, batteries, generator) causes an aetheric flow from a higher pressure/tension/compressed zone to a lower one. Electrons are moved along the potential gradient. We use the electrons to do work, but in the process, we cancel their charge (by means of the positive charge). We can utilise the continuous aetheric flow. In the production of free energy, one merely uses the flow, and not the charges on the particle. The charges will deplete but the aether flow will continue indefinitely if the potential difference is preserved.[7]

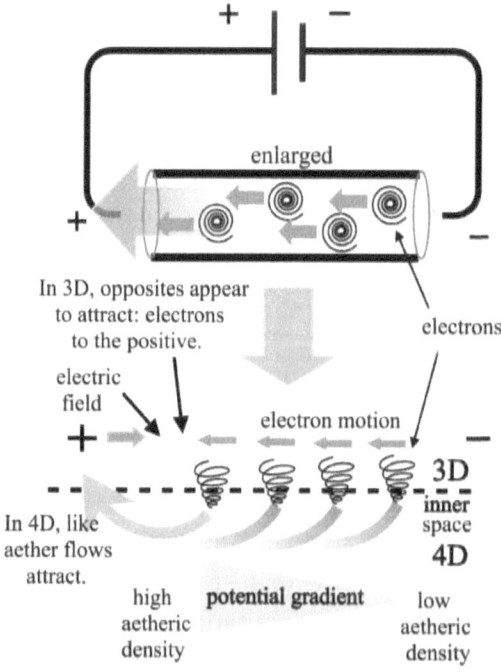

FIGURE 26: Drift of electrons in a circuit and the creation of the electric field.

The reader may still be asking: Why vortices? Why does energy spiral? It is a matter of 4D geometry, 4D oscillations, and transduction of energies from a 4D format to a 3D format. When the spinning effect is travelling linearly, such as electrons in a wire, the spiralling inwards (4D) is still occurring.

In general, the vortex energy must spiral to achieve paths of least action and greatest efficiency and in turn it creates a perfect contextual system—like spheres within spheres—all 'touching', spiralling inwards to lower dimensions (Figure 4). The vortex does not take place in space as though the space is already set up, it formats the basic space structure or aether for different dimensional levels. Clearly the vortex eliminates the dualism (test of truth) of the separateness of space and mass (from which current science suffers) since the vortex has expansion (creating space) and contraction (mass) in the centre.

Consider now the bigger picture of a vortex network on a cosmic scale. See Figure 4, but note that we only show one half of the oscillation (representing our side). If one begins with the universe vortex, one should picture it spiralling in then branching off into lesser smaller vortices representing, say, supercluster galaxy entities. Each of the latter are then visualised as spiralling into further offshoots, which would be galaxy clusters, and from these, galaxies branching off into groups of solar systems, individual solar systems, then planets, atoms, etc.

Thus every particle would be connected to the universe oscillation through the fractal levels, for example, of planets, star systems, galaxies, etc. But remember that each of these spirals meets its opposite (the antiparticle side) to create the stable oscillation or stationary state. We can see from Figure 4 the concept of spheres within spheres, which is basic to all structures (also Appendix D). There is even a similar system of vortices for the tree and human design. The diagram shows that an arm has an energy field vortex at the joints.[8]

Notes
1. Book: *Spacetime and Beyond* by Bob Toben.

2. Book: *The Emerging New Science* vol. 1.

3. This diagram in Figure 23, although limited, is given as probably one of the most efficient as a simple single image for showing the 4D to 3D energy relationship. Later the author discovered that the renowned Russian Philosopher, Ouspensky, described in his book *The Fourth Dimension*, that he had a dream he had discovered the 'secret of the universe'. He awoke and jotted down a diagram. He studied it the next day and included it in his book. It was identical to that in Figure 23, except that he had omitted the spirals and didn't know about vortices. But his analysis was roughly correct regarding the 3D and 4D relationship.

4. Op. cit. *The Emerging New Science* vol. 1.

5. Book: *Engaging the Extraterrestrials: Forbidden History of ET Events, Programmes and Agendas*. Chapter on middle Earth.

6. Under natural conditions particles and antiparticles do not disappear in a matter/antimatter flash of energy, but form a new unity; a higher-frequency particle that can only exist now in the next upper fractal level of higher-frequency spectra, where it divides again into particle and antiparticle appropriate to the new space-time conditions. *The Emerging New Science*.

7. This area has been well researched by T. Bearden. www.cheniere.org.

8. Ibid. Also book: *The Attainment of Superior Physical Abilities*, and articles on skills. www.nhbeyondduality.org.uk.

20.

THE GRAVITATIONAL FIELD

The gravitational field is more complex and deep-rooted than realised; one might have antigravity technologies but still not understand gravity completely.

The phenomenon of gravity still presents us with a puzzling and somewhat abstruse subject. In fact gravity could be considered to be one of the enigmas in science, regarding this mysterious hidden force that causes objects to fall to the ground. Answers have ranged from the early simple view that objects *fall* (no further explanation required), to Newton's gravity law of attraction of bodies, to Einstein's general relativity, in which gravity relates to space curvature (that is, geometry). Finally it is now concluded that gravity not only pervades the universe but is the glue holding it together. As we shall see, none of these is satisfactory and we could go as far as saying that it would be perfectly possible to have a successful technology of antigravity spacecrafts (control over gravity) but still not properly understand gravity.

That gravitational theory is more complex than is generally appreciated is partly due to the fact that one can obtain 3D interpretations which appear to be satisfactory and complete, but they will be found to be dependent on a 4D understanding. At the 3D level, as we shall explain, we may derive a particle or radiation interpretation, but 4th dimensionally the holistic nature needs to be understood, of the vortexual unitary oscillation of, say, the planet, which gives rise to, or reduces to, the 3D and limited explanation.

A particular difficulty is that gravity needs to be compatible with quantum theory and thus must possess quantum attributes itself but at the same time it is expected that it must reduce to the special case of Newton's laws.

Gravity appears to have a single pole; whereas we know from electricity and magnetism that an opposite pole is required to bring about attraction (or repulsion). Also mass only attracts mass; it doesn't repel in our known universe—though we shall see that this is an illusion anyway. Nevertheless, one can't have one pole without the presence of another opposite one.

One can immediately draw certain theoretical tentative conclusions from a scenario in which an object appears to be operating as a single pole (gravity) but its constituents are also involved in two-polarity systems (electricity and magnetism). We can know from 4D geometry that an oscillation (gravity waves) can be at right angles to 3D space; compared with electromagnetic oscillations which act across or within our 3D space, referred to as transverse.

Imagine a taut elastic diaphragm in which vibrations are created by plucking it 1) along its plane, or 2) at right angles to the plane. That is, creating oscillations along the plane or at right angles. We are presenting here a 2-dimensional analogy (the diaphragm) to represent our 3D, which then enables us to see a 4D extension (since it is then seen as 3D). We shall see that gravity is related to the oscillations at right angles to the plane of the diaphragm (a 3D direction); not along the plane. Our electromagnetic waves are transverse and longitudinal and act along the plane (as waves become longer, such as radio waves, they apparently generate a longitudinal component).

In our universe view, only one gravity pole is apparent or detectable in 3D; the other is along a line at right angles to each of the 3D axes, that is, in a 4D direction—this corresponds in the above analogy to a direction at right angles to the elastic diaphragm. The poles for gravity are lined up at 90 degrees to the diaphragm, but for an electromagnetic example the poles are in the plane. That is, with 3D mechanisms both poles are apparent in 3D space, for example, north and south pole of a magnet, or electrons and protons. However, in 4D phenomena, such as gravity, only one pole can exist at a time in 3D.

We might at this point query as to why the gravity system has attraction instead of repulsion (equally probable) since in electricity and magnetism we have both. The reason is: there is no actual attraction. We might mention here that in Newtonian physics it was necessary to have a mysterious attraction concept but this was handled in Einstein's general relativity by space geometry/curvature, though it ignored the polarity problem, and relativity does not give the machinery behind the curvature.

We have covered in the previous section some of the details of the 4D vortex theory. A mass forms at the centre of its vortex. Mass does not cause space to 'pucker', that is, create curvature, as in general relativity. The mass comes into existence simultaneously with the space around it. The vortex is a centripetal spiral transducing energy from 4D to 3D. The centre of the vortex, for example, the mass, is in 3D. The periphery of the vortex is in 4D or higher.

As mentioned in the previous section, these basic vibrations create space-time and mass. There is a spiral oscillation from the periphery (4D) of the vortex to its centre and back. This is a little like imagining space around an object contracting inwards, imploding, and 'condensing' into mass; that is, this implosive, centripetal spiral generates mass. On the opposite, return cycle, the centrifugal action, expands outwards into 3D, dissipates mass and creates space. This action is both 4D and 3D. It cuts across 3D (that is, in the 4D direction) but causes waves along 3D. Figure 27 refers, showing a simplified configuration of dimension, extended in Figure 32.

Thus if one is considering a planet and its gravity, one can envisage an implosive pressure, basically from a 4D configuration into 3D (see Figure 28). This pushes particles and objects towards the centre of the planet. But what about the part of the oscillation cycle in which the pressure is not only released but is 'moving' in the

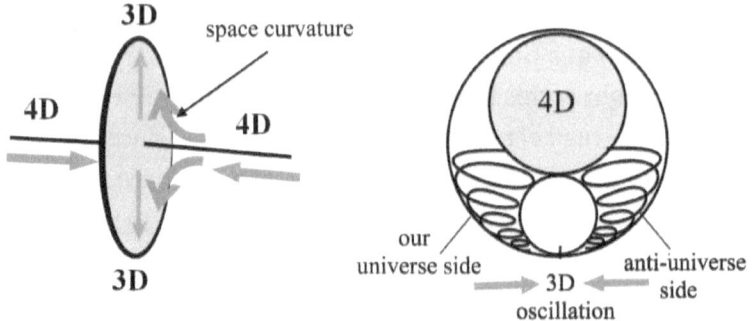

FIGURE 27: Disc represents 3D, showing relationship with 4D flows.

FIGURE 28: For clarity, 3D is projected outside 4D (should be inside).

opposite direction, as antigravity does. We may see that antigravity is the opposite side of the gravity wave; like the opposite side of a coin with both faces identical. For example, the negative of the wave is the same as the positive but turned around (mirror image). Note that in Figure 28 the 3rd dimension is shown outside the 4D circle; it should be in the centre but is difficult to illustrate adequately. The actual shape of the energy is a 4D toroid (doughnut) but one can visualise a 3D toroid. See Figure 32.

At this point let us ask why is there apparent attraction between masses, or more realistically, why do they move together? Take a look at Figure 29. We have two bodies with their overlapping vortices in such proximity so that one is partially inside the 'sphere' of the first, then we can also see that the first mass is inside the sphere of the second mass. In Figure 29 we can see that for mass A the energy is partially blocked by mass B. In fact mass B casts a 'shadow' on mass A. Since the arrows represent pressure then we can see that there is a net pressure from left to right, causing mass A to move towards mass B; and of course vice versa. This pressure is provided by the centripetal action of the vortex. There is a potential difference between the vortex periphery and its centre.

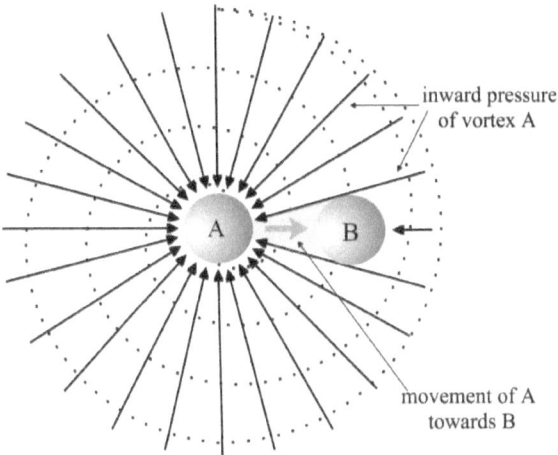

FIGURE 29: Gravitational 'attraction' between two bodies. Here, only body A is shown with its vortex. Same applies to B, which is then drawn towards A.

Now interference from all the vortices in the universe—the planet's vortex, the stars, galaxies, etc.—will create countless nodes, as explained in the previous section. These nodes provide a 3D expression of the implosive pressure to pin objects to the surface, etc. (but as we shall see, they follow the Earth's 4D vortex flow). There is confusion in current physics regarding the relationship between the inertial force, the gravitational force, and the gravitational field.

The gravitational field effect, as described further in the next section on inertia, is quite different from the inertial force and gravitational force. The principle of equivalence in general relativity applies to the inertial and gravitational *force*. However, there is not actually any principle of equivalence because these two forces are identical; that is, they are the same. Gravitational nodes passing through a body resting on the Earth's surface are constantly trying to drag it downwards as one set of resonating nodes after another try to lock onto the nodes within the body. This is the same effect as a body out in space which is, say, being pulled by an artificial force, causing acceleration of the body and the dragging of the body's nodes away from space-time nodes, which are attempting to lock onto one another by resonance. See next section.

This is the Newtonian force. However, there is no Newtonian force when the body is falling in a gravitational field; it is carried by and resonates with the space-time nodes (free fall; goes with the flow).

So far, we have only described the effects of the vortex which expresses gravity in a 3rd-dimensional sense. That is, the nodes, which we may call gravitons, press towards the centre of the planet and pin objects to the surface. This is the gravitational force, and is identical to the inertial force. The vortex is an oscillation; it is a centripetal energy spiralling from a 4th-dimensional to a 3rd-dimensional configuration. We shall now attempt to explain this more basic property in Figure 30.

One can see that this is the basic vortex, as illustrated in Figure 30, now shown horizontal but again omitting the 4D connection, that is, the ends A and A' are not shown coming from the same source (4D) for convenience of illustration.

Let us recap on the basic 4D dual vortex description. One side of the oscillation appears in our 3D space, that is, one pole. The other pole, or flow, is antigravity relative to our gravity (they are both gravity in opposition). Note that the following description will present initial difficulties, but keep in mind that it does explain, and thus supports the fact, that some physicists have concluded space has both positive and negative mini-blackholes and mini-whiteholes. That is, four types: a negative ingoing blackhole and a positive ingoing blackhole, and similarly for the whiteholes. This can have some meaning when we consider that the gravity field has a hierarchy from atoms to planets, stars, galaxies, etc. all superimposed and interfering, creating infinite minute vortices/nodes/-gravitons of space. With this in mind, one has to imagine this topology structure incorporating the above four-polarity system, giving four different 'hole' attributes.

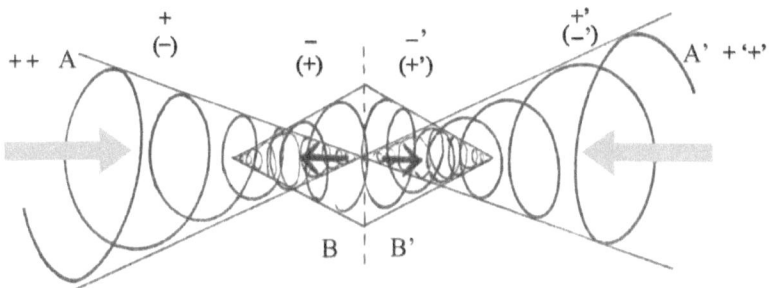

FIGURE 30: Gravitational vortex

In this Figure 30 we see that there are polarities A to A', a 4D oscillation, but within this are A to B and A' to B' which are 4D-to-3D oscillations. These latter two would correspond to Dr. Russell's centrifugal components (the energy is coming out into 3D). Thus on the left of the figure, say, our universe side, we have the 4D, A - A' flow which is at right angles to our 3D. Within this we have A - B, a 4D-to-3D flow providing the 3D gravity aspect of particles pressured towards the centre of the Earth. The same applies mirror-imagewise on the anti-universe side. However, only the 4D flow cuts all the way across from A to A'. Thus the 3D anti-flow of particles from the anti-universe don't affect our side. Nevertheless the 4D comes in on the anti-cycle into our universe side—at 90 degrees to 3D.

It is only the 4D aspect which comes through on the anti-cycle and it affects the particles in our space differently now. Instead of creating changing potential in space as does the gravity half cycle, via gradient 4D-3D, it influences the particles to have the same potential across space—this can be called waving in time (this is a scalar field). It thus gives a rectification of the particles. These gravitons pressure towards the centre of the Earth on the gravity half cycle but don't reverse on the anti-half cycle, since the spatial effect on them has changed to temporal. There is a reciprocation of time and space—see Figure 31. This is a theory which has at least the potential to explain what is happening with these complex and hidden processes.

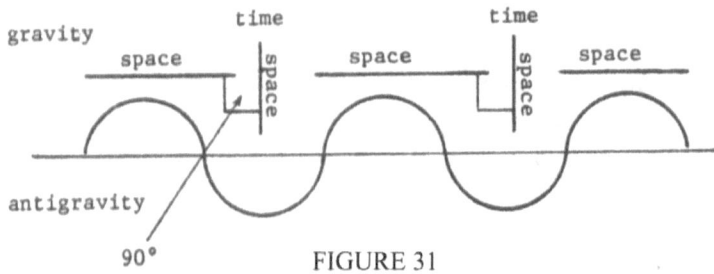

FIGURE 31

Continuing with this theory, and at the risk of repetition, recall again that Figure 30, is a 4D view of the same configuration as Figure 28 (or Figure 32), except that A and A' are not brought together and emanating from the same 4D source. The extra vortices in the middle are due to an inductance that occurs in 3D due to this oscillating 4D gravity/antigravity wave. This is a transformer-like action which creates a hysteresis effect. Without this hysteresis effect there would be no 3D universe as we know it. What this effect does is to create not only separate 3D vortices B and B' (on each side) but vortices which are in a different phase state with their originating 4D vortex. There is a hysteresis lag of the 3D electromagnetism B and B' induced by the gravity/antigravity oscillation. The phase state of such 3D vortices would be expected to vary for different vortices. Thus compared with Figure 28, Figure 30 now includes the separate 3D secondary vortices.

As the 4D energy from A (or A') pressures inwards and transforms down into the 3D matrix at B/B', the energy builds up at B/B' similar to the action of a condenser, in which 180 degrees later, as the cycle of oscillation AA' manifests A as dominant, B discharges into our 3D in this 4D A-flow. Thus B represents a centrifugal outgoing component of electromagnetic energy into the 3rd dimension. It results from the hysteresis effect of the 4D oscillations spiralling into 3D—the fundamental oscillations experi-ence a dragging effect similar to the effect of electrons oscillating in an antenna, causing the fields to interfere and propagate outwards producing electromagnetic waves. Now this happens on both sides of the centripetal oscillations, and would be expected to occur with all continuous or standing-wave vortices. Figure 32 shows the secondary vortex from a 3D (more realistic) viewpoint.

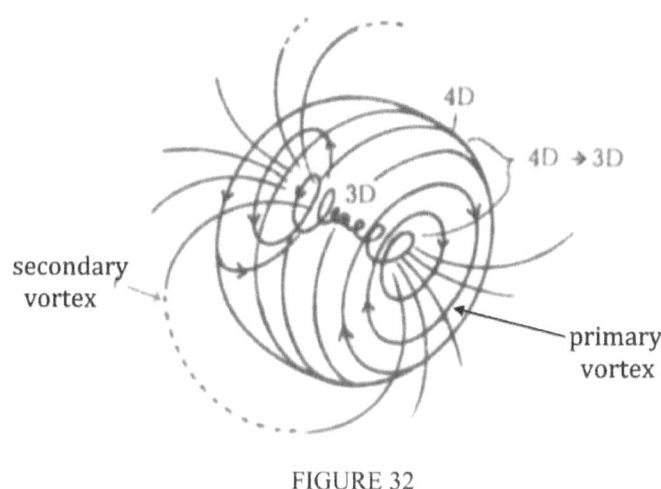

FIGURE 32

Let us summarise this difficult material. Figure 30 shows the fundamental and dynamic effect of what causes gravity in isolation. On the left side A represents the 4D centripetal gravity flow in our 3D. On the right, in the anti-universe side we have an identical situation but in reverse, that is, mirror image. (Note that just as plus (+) and minus (-) represent opposites, the different type of opposite is given by the use of the apostrophe (').) This nevertheless gives us a potential for antigravity on our side since it is basically all superimposed. The difficult geometry to visualise is that A and A' are from the same source (see Figure 26 and 30) but each opposite side with respect to the other is 'inside' the other, not extended along linear 3D space. The centripetal spiral A pressures in and by hysteresis and inductance creates the central electromagnetic centrifugal spiral[1] and expands outwards into our 3D.

We might note that we have two 4D poles within which two 3D poles (on each side) are formed. This could be referred to as a four-polarity system. This configuration would apply to all natural entities from atoms to planets, stars, galaxies, etc. with a resultant infinite interference effect creating a myriad mini-blackholes and mini-whiteholes of space-time.

The primary 4D gravity/antigravity vortex would be expected to create a secondary out-of-phase radiation vortex due to the hysteresis effect. The gravity component and the radiation component form our 3D space (and the antigravity and its radiation component form the space on the anti-universe side). The 4D antigravity potential on our side is homogeneous spatially but varies temporally and is not normally detected, nor does it subtract from our gravity; similarly for the other side.

Thus the primary gravity vortex induces a secondary electromagnetic vortex with lines of electromagnetic force that are at right angles to the centripetal spiral (lines). These primary and secondary vortices interfere to produce a consistent grid system around different entities; described in the previous section. The points of interference create criss-crossing lines of force, forming nodes in repeated patterns of motion (for example, orbits) for establishing, say, electrons in atoms, or large nodes around the Sun for carrying planets. See Figure 32.

Gravity thus appears to have a dynamic wave phase and also a particle phase. The dynamic action is the oscillation caused by the macro-polarity, 4D to 3D, pressuring centripetally into the 3rd dimension. The static phase is the effect of all the minute interference nodes and gravitons passing through objects as they are pressured implosively (by the dynamic phase) towards the centre of the body, such as Earth.

The New Science

Now let us look at some different aspects of gravity. Is gravity a force-field? Science assumes it is, since if we place a body in the gravitational field it is acted on by a force, to pull (actually push) it down. (Exactly the same applies to the charged particle, such as an electron placed in an electric field.)

In the New Science the gravitational field is a scalar field. It is in a state of quantum action. It is not a force-field. The fact that gravity appears to have one pole, should tell us that since there must always be two poles, the other one (the antigravity side) must be hidden from 3D, and that it is along a 4D direction. It is the other pole of a 4D vortex. In other words, the gravitational field will be found to be a 4D scalar field. This has compatibility with quantum physics, that is, quantum action, E x t (energy times time).

Thus objects in free fall in space are in a phase-correlation condition with space-time. They are observable as per the first degree quantum reduction. They are in harmony with the universe (their frequency patterns are coherent—object and universe as a one-bodied system). Immediately we interact with such a body by pushing it, slowing it down, or using a two-bodied propulsion system, such as a rocket, Newtonian laws then become applicable: forces, momentum, mass, kinetic energy.

The body is now out of phase with space-time; a final macro-step of phase randomisation, and is in the condition of a two-bodied system. It is disconnected from space.

As far as technology is concerned what is the difference? In the two-bodied system, or second-order reduction, the body is being moved by surface interactions, via a 2D-to-3D (dimensional) interface. This interface is where the impulse, causing motion, contacts the body structure.

In the first quantum condition, free fall, the impulse and body have a 4D to 3D interface (note the 4D field spectrum acts as cause). There is complete penetration of all atoms down to their nuclei of the body by the gravitation field (or artificially applied scalar field), entraining the atomic oscillations, resulting in no inertia, or mass (or kinetic energy) being invoked. Obviously as soon as kinetic energy is measured (and the body's motion is interfered with), instantly the 4D - 3D interaction changes to a 2D - 3D interface, and Newton's laws are then applicable.

Mechanical systems involve physical interaction of the operating parts. This will always give us the second-order quantum reduction and Newton's laws are obeyed.[2]

To avoid the second coherent reduction, field systems must be used that act fourth dimensionally—not magnetic, of which the latter merely penetrates 3D-wise and thus still only invokes a 2D - 3D interaction. Scalar fields will act 4D-wise to satisfy the condition of avoiding the second quantum reduction and utilising the coherent state following the first quantisation. This gives the necessary coherent state, which naturally manifests after the first observational process (which brings about probability selection from the first quantum reduction).

Thus in the New Science, quantum vacuum is basically a scalar field, which has electrostatic potential, that is, a state without mass: real particles, photons, electrons (the actual Absolute described earlier would underlie this). The force property of the field only arises when mass is present, and even then such particles need to be constrained or artificially moved externally in order to 'express' their Newtonian characteristics: mass, force, inertia, kinetic energy.

This is not recognised by mainstream scientific thought. A scalar field is not a force-field, as is the electric field, unless mass or real particles are present, and even then such particles need to be constrained or acted upon artificially, or by random forces, and moved to invoke the force (but there is no force in 'free fall'). The scalar field is a coherent field state (as is indicated by quantum-field theory itself). See Appendix E. Similarly, as we mentioned earlier, a body in the gravitational field in free fall is not subject to the gravitational force—it is coherent with the flowing space-time gravity pressure. The gravitational field is not acting on the mass with a (separate, external) force—which would be a decoherent condition—until the free-falling body's motion is arrested externally, or when pushing it, resisting it, or letting it rest on the ground. Note that Einstein recognised a falling body was weightless, in particular, when we experience personally jumping from a height. Unfortunately he allowed the unbending authority of Newton's laws to dominate (force merely cancelled by inertia at the acceleration due to gravity) and missed the clue that the body was coherent with space time.

As a final comment let us emphasise again the potential similarity of this model to the ocean analogy in which all of creation is nothing more than patterns and curvatures in the infinite quantum realm. In the view of the author, this abrupt change from the quantum realm to the material world, which science teaches, is fiction. There is not just the primary collapse of the wave packet, from the supposed coherent state to decoherence, but secondary quantum reductions in a whole fractal hierarchy, which we dealt with in volume 1 (Figure 38, Appendix D).

Does anyone really think that a truck swept into the air by a tornado and steadily held in suspension is achieved by air pressure only? The suction effect is accompanied by anti-gravitational vector —but remember this is one of the heretical words and any professor at a university can be dismissed for researching anti-gravity.

Notes
1. www.nhbeyondduality.org.uk. Article: *The Basic Energy Unit*. Also book: *The Secret of Light* by Walter Russell.
2. Book: *The Emerging New Science*.

21.

INERTIA, A COVARIANT AETHER AND THE VELOCITY OF LIGHT

Science's method of solving the puzzle of the covariant aether was to discard it.

Let us recap thoroughly on previous related information to assist in a better understanding of inertia, a covariant aether, and light velocity. The New Science utilises the ocean model that allows us to eliminate particles as separate entities—one can't pluck out the ocean whorl (tiny whirlpool, characteristic of an electron); it is part of the ocean. Recall, this is supported by quantum physics: All bodies and objects are like shapes and variations in the structure of space (Schrodinger).

Now as we proceed deeper into structure of matter using the vortex model, we similarly have the formation of orbital electrons; particles that appear to have a separate existence but manifest from intersection of vortex lines of force. A shell structure forms naturally from a dimensional analysis of the dual vortex. If we recognise that the two-polarity vortex system expresses first, higher dimensionally, we can find that just as two intersecting 1D lines form a point of zero dimension (one less), and two intersecting 2D surfaces interact to form a *one*-dimensional line (one less), so two 4D intersecting flows will create a 3D 'solid' spherical effect, and for coherent waves the interference pattern will produce the characteristic light and dark fringes—alternating light shells with dark shells in which the electrons appear to orbit (in the light shells).

These node characteristics, caused by intersections, are of particular interest, especially the atom-sized ones.

In addition, however, a planet's vortex system will similarly cause a huge variety of intersection nodes of its spinning electromagnetic lines, ranging from large nodes for anchoring potential satellites, to a ley-line grid system. These are like countless intersecting ripples on water. By extending this principle throughout space we find similar intersections of waves on all dimensional scales caused by stars, planets, galaxies and universes, etc. (Figure 21). There will thus not only be major nodes caused by these large bodies but infinite minor nodes giving rise to the so-called mini-black and mini-white holes, coming and going fleetingly—essentially virtual particles. Recollect that these intersection points cause oscillations at right angles to 3D space, creating small apertures in the aether fabric, allowing 4D energy to enter 3D (hence the term mini-blackholes and mini-whiteholes).

This would be the original superspace: a quantum turbulent foam of particles and waves. This means then that there are particles with *all* velocities and directions passing through every point of space, possibly including all frequencies and accelerations over a certain range. This sea of infinite particles forms the basis of both our familiar but mysterious property, inertia, and also less familiar, covariant space; a requirement for a proper theory of the aether.

Note that strictly, these particles are nodes and can pass through one another. As we have indicated, these are the nodal regions of (4D) oscillations, cutting across the 3D matrix originating from the countless higher-dimensional vortices as they fragment into smaller and smaller entities and interfere with one another. (To recap, note that two 4D beams projected holographically from a 4D configuration into 3D, where they interfere, create a stationary state which can appear to us as a solid spherical particle—this argument is just 4D geometry.)

Thus we have a model of space as composed of infinite nodes covering all possibilities of direction, velocity, frequency, acceleration. As alluded to, shortly, we shall see that this is a prerequisite to an understanding of inertia. All possibilities are covered by these nodes, for instance, at any point in space, particle nodes pass through in all directions, and for any one direction a complete range of nodes exist for velocities and frequencies.

Thus this framework is necessary for a complete description of inertia, which we shall find is contextual and of an electromagnetic nature, meaning it is dependent on what kind of an interaction is impelling the body.

Scientists and laymen alike have been programmed with the notion that objects *possess* inertia—as though this was some absolute inherent property of mass. If one really contemplates this and breaks down the programmed thinking, one will see how erroneous it is. What is this strange ingredient (inertia) with which we attribute mass?

Imagine a heavy object flying through the air, say, an airliner. We automatically invest this body with ponderousness, weightiness, irrespective of what the body is doing or what is happening to it. If a body is in free fall, that is, falling freely under gravity or moving freely out in space, or in orbit (also a projectile after firing is included here), there is no inertia manifesting, not even momentum (mass times velocity) or kinetic energy, etc. But one will insist that the body has velocity and mass, and therefore momentum. It has velocity but the mass in momentum is based on gravitational mass, which is identical to inertial mass, and as we shall see there is no innate inertia.

Amazingly we have said nothing that violates Newton's laws (where they apply), and every educated person will agree with the above when we remind them that inertia only arises, or is detected, when we interfere with the motion of the body. The above presentation was to indicate the preconceived and programmed state to which our minds become prey; that is, that an object possesses inertia. It is necessary to deprogram this false thinking before proceeding with correct thinking (that is, imagine all bodies

The New Science

as a mass of oscillating fields with no inherent properties of inertia—think of the object as, say, being 'lighter' than a feather).

Let us take another example. A huge spacecraft, say, weighing 10 billion tons is lighter than a feather relative to its propulsion system (assuming it is operating 100 percent efficiently), but if one pushes the craft with a Newtonian force—even while the propulsion system is operating—it will resist with a force up to 10 billion tons. Advanced systems do not have to cancel inertia, they bypass its onset.

Now let us explain how inertia arises. Keep in mind the phenomenon of resonance and, in particular, that if two oscillating particles come close that have the same frequency they will pull together and lock on. Also recall that a body moving freely through space is in what is referred to as free fall. Free fall also includes an object coasting upwards (after the impelling force has been released), and similarly for all bodies in orbits, that is, satellites, planets, stars, etc. It doesn't matter what velocity it has relative to anything or what direction it has, there will be nodes of space-time with the same direction and velocity, and also of the same frequency as the nodes of the structure of the body (for example, the centre of an atom is a node, and surrounding it are many minor nodes).

Thus these space-time nodes, penetrating the body and resonating with the body's own nodes, are in holistic relationship. The object could be said to be 'at one' with space-time. There are no forces, no inertia, no momentum, etc. The body is in a coherent state with space.

Now consider contacting the body at its surface, resisting its motion. Let us just explain that the space-time nodes are moving as a result of the electromagnetic or quantum potential gradients of space. The nodes endeavour to keep the body moving at their velocity and in their direction since the space-time nodes are locked on to the body's own nodes by resonance and entrainment. It doesn't matter which direction the body is going in, or what the velocity is, there will always be nodes moving correspondingly to the body's behaviour. Thus the body appears to resist any attempt one might make to alter its motion, owing to the attraction of the nodes. If we succeed in reducing the body's speed, even a minute fraction, the body's nodes will now have locked onto another set of space-time nodes which correspond to the body's new velocity, direction, and frequencies. Thus while we are successfully resisting the body's motion we cause the nodes of the body to be dragged away from the corresponding space-time nodes continuously—causing the body to move from one set of space-time nodes to another. Clearly this also applies to any manner of Newtonian interactions for deceleration, acceleration, and deflecting the body from its path, etc. These circumstances will still apply in a gravitational field where nodes are under acceleration and deceleration, but it is a little more complex.

We see from the above analysis that a falling object does not obey Newton's laws. Only when the object is resisted in its free fall, such as impacting Earth, does it obey Newton's laws. When it is falling, it has acceleration without force. It is in holistic, coherent equilibrium with its environment. The gravitational field is scalar, and not a force-field.

If the body is restrained from falling, such as when it rests on the ground, the space-time nodes (which are being pressured towards the centre of the Earth—we can call them gravitational nodes)—will pass through the stationary body on the Earth's surface. Many of these nodes of identical frequency, passing close by and through corresponding ones of the body, will attempt to lock on but continue through, causing a dragging-down effect.

Thus one can see clearly that this is an inertial effect but which is called a gravitational force (this equivalence of gravity and inertia agrees with orthodox science). The body is constantly being dragged downwards but is prevented from going with the flow by surface contact. This condition of being pushed against the ground with nodes temporarily hooking together and then being pulled away is entirely different from that of the body in free fall, where space and body flow together, and this gives rise to the well-known weightless condition.

What we are stating here is that when the body is falling freely there is no actual gravitational *force*. The body is not being pulled down by a Newtonian force as described by science (Newtonian physics), that is, a gravitational force, causing acceleration and in turn causing an inertial opposite force. The same space-time nodes entrained with the body's nodes remain in a state of coherence throughout the free fall period; they are not being pulled away.

Thus there is no inertia. There is acceleration without force. This holistic condition superficially appears to be the same as the Newtonian interpretation in which the object is pulled down by a gravitational force and the body resists with inertia so that the gravitational *force* and *inertia* are equal and opposite at the acceleration due to gravity.

This is a typical illusion in coherent systems. Mathematically they cancel out. But a true no-force condition is *not* the same as one with cancelled forces—it involves a completely different science. There is no gravitational force and inertia in the first place. So naturally *force* minus *inertia* (f − I = 0) equals the same as in our coherent case in which force and inertia are not present, that is, are zero. Thus in the orthodox case, an illusion is created (actually an artificial construct) that force and inertia are factors which are in counteraction. This is an excellent example of the assumptions that science makes and overlooks (that force and inertia still exist), when they are not present at all (in our New Science coherent systems).

We see then that if we apply a Newtonian force (2-to-3 dimensional interface) to a body, we invoke the inertial property. If, however, we apply quantum action (4D action) or more specifically a 4D-to-3D action, no inertia will be generated. The onset of inertia is avoided—it is bypassed.

The gravitational field, or an antigravitational field (or scalar/quantum fields) satisfies this condition. They act 4th-to-3rd dimensionally on every nuclei simultaneously and there is no reaction, no resistance, no inertial effect whatsoever. We thus see that inertia is contextual; its onset depends on the interface one adopts in the interaction.

Now this theory of a superspace of infinite complexity, containing particle and wave interactions, not only forms the basis for an aether but one very different from the aether envisaged by scientists. After this concept of the aether was introduced, it was then abandoned because, for one reason, the theory didn't possess covariance relative to all motion of bodies. However, this new superspace can form the basis for a covariant aether. Why exactly did this need arise for a covariant aether?

It is considered by many researchers that the aether has (is) an absolute frame of reference. But also, and in particular, by others, that no one knows what an absolute frame of reference is. As a result, its current understanding is merely philosophical and not scientific. Physical properties have been assigned to the aether but not fully accepted and no experiment has clarified the existence of the aether. Clearly the aether needs to be upgraded from a philosophical concept to a scientific one.

The problem that arose, however, was regarding measurements of the aether, for example, could we detect the motion of the Earth through the aether medium? The aether was considered to be in some sort of absolute state, at rest, or in a fixed condition in the universe. Thus it was expected that an aether drift would be detectable around our orbiting Earth. The covariance would still be relative to field and energy systems; that is, particles/waves, and potentially quantifiable; this is not the true absolute underlying this.

The celebrated Michelson-Morley experiment was conducted in an attempt to detect the aether drift relative to Earth. However, none was measured, and this also meant the aether itself wasn't detected.* In spite of this, the notion of aether drift nevertheless leads us into a weak physics premise. The fact that there was apparently no change in velocity between the two directions not only gives us the possible solution that the aether can't be detected (or is not there) but also that there is true covariance, and that is why the aether couldn't be detected. In fact, even this argument isn't required. We know from our vortex theory that each body, such as Earth, has its own dual counter-rotating vortices, and since these create the body of the Earth, they remain around the Earth. Therefore again no change in the velocity of light would be detected; the light detecting apparatus moves with the Earth and its vortex. [*A wave on water will appear to travel faster towards an oncoming boat than when the boat is moving away, viewed from the boat.]

The idea of, say, the aether being at rest relative to one body, then having different motions relative to every other body moving relative to the first body is questioned by the 'tests of truth' in physics. The correct logic would be that the aether would need to appear the same for different bodies moving at different velocities, etc. A proper aether theory must have equal applicability to all bodies no matter what their motion is. Such an aether medium would be covariant—meaning as the body's velocity and direction vary through this aether it reveals co-variance by presenting a different group of particles, corresponding to the body's motion but always a group which shows no change in its relationship to the body.

The Michelson-Morley experiment also showed that the velocity of light did not apparently have a variable relative velocity. If the aether drift—due to the Earth moving through the medium—had been detected, then the light which was propagated through the medium would be expected to have had a velocity which would be observed to vary according to which direction the aether drifted relative to the measuring device

The fact that the velocity of light was found to be constant, implying then that the aether was not detected gave Einstein the opportunity to ignore the aether. Einstein claimed that the concepts of ether and absolute reference frame were unnecessary, and he used those same transformations in order to save the principle of relativity. This was quite ironical and contradictory in a way, since if there were true covariance of the aether then by the methods used and the apparent constancy of the velocity of light we can surmise the aether wouldn't be detected. That is, the attempt to detect the aether by an observed change in the velocity of light and not detecting this, supports the notion of absolute covariance (of the aether); however, at the same time it created an opportunity to ignore it. Ironically, if the measurement had been successful it would mean physics would have had to settle for a relative background medium (space), which would be associated with a limited locality (not Absolute but could be large, such as a galaxy). Not a good solution. This is quite low on the physics 'tests of truth' scale. If there had been a proper covariant theory of the aether this could not *only* have explained the constancy of the velocity of light (and put it in proper perspective as a source of variable velocity) but it might have been recognised that this constancy was an illusion.

Nevertheless physicists were happy to drop the notion of this mysterious medium called aether. Little was it understood that this decision to discard the aether notion put the lid, so to speak, on the 3rd dimension, and another (limited) physics datum was established to arrest true advancements (which means not just in physics but the effects of knowledge on evolution). In particular, it caused a lack of interest in our most important physics model; the ocean model in which wave patterns, vortices, or whorls, were seen as particles, bodies, objects, and manifestation generally. As we have stated previously this was a big mistake.

Einstein's special theory of relativity, which postulates that the aether cannot be detected and that the velocity of light is always constant relative to an observer, is a mathematical contrivance to handle why different observers travelling at different speeds relative to, say, a beam of light will all measure the same value for the light velocity. This manipulation resulted in the by-passing of the real nature of the aether and the potential to understand it.

We have already provided the ingredients to model a covariant aether: the infinite particles in space, travelling in all directions, at all velocities up to that of light, and possessing all frequencies (within a particular spectrum range). If one imagines moving through such an aether in any direction with any velocity the relative properties of the adjacent aether are unchanged owing to the complete range of node characteristics described above to match any observer change. Similarly, light is propagated according to the aether properties. The observer only perceives visible light carried by nodes with velocity characteristics relative to the observer. Another observer moving at a different velocity will perceive another range of nodes that has the same *relative* velocity; and thus observes the same value of the velocity of light (gauge invariance). The reader may find difficulty with this concept of how a covariant aether can be explained and also how the velocity of light can appear to be always constant. An example of what is called gauge invariance may help.

Imagine a voltage outlet with a scale of values ranging from zero to 1000 volts. Also that there are terminals at voltage points: 5 volts, 10 volts, 15 volts, all the way to 1000 volts. This scale of changing voltage values can be an analogy for the aether variations we mentioned, that is, there is a whole range of node (particle) velocities, directions, etc., from every point. Keep in mind that the aether's interactions with various bodies, are such that the nodes are required to 'see' the aether as covariant (the same for all these bodies).

Now back to the voltage analogy. We require next, a potential difference to, say, run an appliance, such as a small motor. We can connect the terminals of the motor, say, with crocodile clips to any adjacent pair of voltage terminals on the voltage scale: say, 0 to 5, 5 to 10, or 50 to 55, or 995 to 1000. The motor receives each time the 5 volts; the potential difference. The 5 volts can be tapped off at any two adjacent terminals and, in effect, the motor can't tell any difference. Similarly the aether, to have a valid theory, must present the 'same face' to every object present and active in the aether.

This may also help an understanding of why the velocity of light always appears constant whatever the relative speed between the source and detection device.

In addition to the effect mentioned above with the vortex model we don't require that the Earth drags the aether. The Earth, as with any other natural entity, is the centre of its vortex. Obviously the vortex goes with it and it is the vortices of space which carry the vibrations. Thus the moving planet takes the encompassing vortex with it, and this would be expected to obscure any predicted change in light's velocity.

It should, however, be added that aberration from star light is observed to occur (an academic point), indicating a lack of aether drag by the Earth. But in the vortex theory the vortex is oscillating and one may still observe the aberration and still measure a constant velocity for light using the Michelson and Morley experiment. To make this a little clearer we have 1) the Earth supposedly dragging a dense medium, aether, like air, compared with 2) the Earth, being carried by its vortex (not at all like air), of which the vortex is oscillating in the aether medium, and then we have light being carried by the vortex oscillations. Thus (1) is not required. And yes, the velocity of light would vary as it propagates through the vortices of space in the universe (it will also create a red shift owing to acceleration into a vortex, lengthening the wave).

Keep in mind the big picture of space, the full strata of higher-dimensional levels of the holographic fractal universe, providing endless underlying inner-space nested levels of energy interactions and subsequent nodes. As we go within, into this vast matrix of space-time, we ultimately arrive at the Absolute which is all consciousness (in many different energy formats) and basically can be considered to be in a virtual condition. Any energy though is inherently present to be rekindled, regenerated, that is, brought into materialisation as per the infinite possibilities of the quantum realm. In addition to the chaotic turbulence of quantum foam, there will be a whole host of momentary random quantum regenerations. For example, separate 3D particles of the 3rd dimension (or, say, light waves, similar to forming a laser) can be correlated—their frequencies made coherent (in resonance)—so that their combination forms a new whole (section on vortices). This is not a composite of parts but a holistic, quantum state (it can also be called collective). This new quantum state will have a higher frequency than its parts, and also its wholeness will be greater than the sum of its parts. Thus even coherent states will emerge randomly, but amongst the ordered (coherent) fractals there would also be chaotic fractals.

For space, however, many of these discrete vortices should be considered as in a virtual condition. Nevertheless materialised vortices, such as those for galaxies, stars, planets, etc. will, as previously stated, all interfere with one another regarding their emanating waves, causing infinite sub-vortices and nodes within space, and subsequently countless grid systems. The diagram in Figure 4 does not give all the manifested fractal levels for our universe, such as clusters of galaxies (a whole entity), molecules, etc.

Thus this superspace is a virtual-state hierarchy of infinite contextual and potential fractal levels on a fine gradient from 3D into higher dimensions from which all creation manifests; and it is, of course, also zero-point energy, the basis of free-energy devices.

Light travelling great distances will pass through many different vortices: galaxy, stars, planets, etc. While in the vortex, the light will travel normally instantaneously. But note that there will be a delay as it jumps from one vortex to another of different frequency. Similarly one would expect to find that advanced space crafts, which are capable of teleportation, may take a few minutes in transition; the time being taken to switch frequencies from one vortex to another. Thus we can see that evaluations of light from distant stars, revealing the past, based on normal light speed, will be flawed.

If one conducted the same experiment on Earth, that is, locally, owing to the relative velocities tested between the observer and observed, we again may find the appearance of the constancy of the velocity of light. Recall the fractal gradient underlying all creation, as applied to the velocity of light, it would mean light had inherent within it all velocities; relative speeds will alter the frequencies received, causing selection from a different range of frequencies from this light source, covariantly compensating and resulting in no change in the constancy of light's velocity.

It was concluded that the velocity of light was always constant no matter what the speed of the observer was with the measuring device; that is, the light velocity is considered invariant with respect to any (moving) reference frame but covariant with respect to the aether. The latter is not a true absolute; it mimics the Absolute, energy-wise. The Absolute is beyond all particles, waves, space and time and is 'everywhere at once', a true zero or stillness of motion simply because its nature is absent any property that is related to space and time. But with the vortex explanation light velocity does not have to be constant. If this *was* the case it would have a drastic effect on a large section of physics. The measured value of the light velocity will always appear to be constant if the aether vortex is creating the matter of the body with the body at the centre of the vortex. Out in space the velocity would vary. Once science establishes the (illusions of) velocity of light as universally fixed then appropriate distortions of space, time and mass have to be contrived to preserve the constancy of the light velocity. Then the third-dimensional limitations of the scientific instruments will confirm this.

By fixating the light velocity as a fundamental constant, it acts like the fulcrum of, say, a see-saw—a zero point from which measurements are referenced. This zero point is relative to the third dimension and the 3D frequency spectrum or rate of information, or *rate of creation of space*. But our 3D is 'carried' by the next fractal level of space-time (which will have hyperspace properties) and a new 'zero'.

To go into more detail on Einstein's special theory of relativity a new concept needs to be introduced; that of rate of creation of space (and matter)—different dimensions have different rates of information. With this, the illusions of the special theory become apparent, in particular, the falsity of the relativistic increase in mass as the velocity increases.

Regarding this relativistic mass-increase, we are now interested in the mass-distortion illusion of matter when approaching the speed of light, which we shall see is clarified by the rate of creation of space. Relativity theory indicated that distortion effects of a body will occur at high speed, approaching light velocity, and that at the theoretical speed of light the body's mass will become infinite. Let's analyse this.

Special relativity says that as we push a body close to the speed of light its mass increases rapidly in the higher range up to infinite mass at the light velocity. Now we must admit that particles have since been accelerated by magnetic fields up to close to the light velocity and the force required to do this is known to increase exponentially, which projects to infinity at the light velocity. But let's look at the assumptions here. Yes, the force applied is increasing as per relativity, so therefore the resistance is also increasing proportionately. Everything is fine at this point. But we might ask, is this resistance the normal one that has increased exponentially, or is it an extra resistance factor, which has emerged? Science assumes it is the normal resistance due to inertia—a big, unjustified assumption.

Let us review this. Firstly, we have motion of the body with accompanying inertial effects as expected. As we have seen, in this condition, the body's atoms are non-coherent amongst themselves but are in local resonance with the mini-nodes of space as the body accelerates which causes resonant switching of particle nodes, creating the effect of inertia; a familiar condition. However, at very high velocity, the body's linear speed becomes of the order of magnitude of the motion involved in the space oscillations. In effect, the body has to wait for the space to be created to move into it. But science has determined that the inertia increases exponentially; that being so, we deduce that this means the mass has increased likewise.

This is the thinking. Science recognises that our material existence is in 3D, and that these three dimensions are orthogonal (at right angles to each other). However, there is no detection by science of the spectrum of frequencies that these dimensions encompass. Scientific instruments can't detect them for the simple reason that the instruments themselves and the brains and physical senses of the observers are manifestations from the same frequency spectrum (a machine can't evaluate a machine of the same order). Recapping yet again, the vortices of all bodies, such as planets, stars, galaxies, are creating space-time as well as mass, and their waves interfere, causing criss-crossing of energy lines and a myriad of mini-black and mini-white holes, appearing as micro-particles and virtual particles; nodes, in fact. There are thus countless vortices, large and small, forming space; it is a mass of oscillations, in effect, primary and secondary nodes. An oscillation has a frequency 'swinging' from positive through neutral (zero value) to negative. Thus a basic oscillation is like a pulse switching on and off.

Keeping the above in mind, we can now see what happens when an object is accelerated *artificially* within this field of activity (this is not the same as free fall acceleration). Note that as a whole it is out of phase with the space-time nodes, that is, its own oscillations (of atoms) are not coherent with space-time. Now when travelling at very high velocity, if its linear speed becomes of the order of magnitude of the rate of motion of the space oscillations, it is held back during the 'off' state of the spatial oscillations. That is, it has to 'wait' for the space to be created in front of it. There is thus a dragging effect on its motion, characteristic of resistance. Since these first-order space oscillations also determine the velocity of light, the body will meet with impossible 'resistance' at the speed of light.

This would be the additional 'resistance' (to the inertia) that impedes the body through space as it nears the velocity of light. It is the relatively low rate of information of 3D space that prevents the body going faster. But this will be a 3D fractal level caused by the mini-oscillations (vortices). There will also be larger oscillations from larger entities, such as planets, stars, galaxy. These are greater coherent states/nodes with higher frequency, giving rise to inner layers of hyperspace underlying the smaller nodes. If the body's complete structure itself is made coherent (entrainment of atomic oscillations that raises its frequency) it could theoretically match and resonate with this next (or higher order) hyperspace fractal level of higher frequency and then be subjected to a new and higher upper limit to the light velocity—meaning a greater speed of light in higher (hyper) space. And so on to the next level. However, this body is still obeying Newton's laws but is only experiencing normal resistance now due to genuine inertia alone and an absence of relativistic mass.

In the Michelson and Morley experiment we are looking at a global example in which an observation is being made from Earth of the light from a distant star; a motional relationship between the Earth and the star. Light enters the vortex of Earth from beyond Earth. This does not require any covariant theory since a velocity change is prevented by Earth carrying the vortex (aether). As already stated the vortex moves with, and creates, the Earth. So there cannot be a detection of any relative motion of an aether. However, in local observations of, say, an artificial light source on Earth, a change in velocity brought about by the measuring equipment moving towards and away from the light source would be expected to occur. Thus a covariant explanation is required, as given above.

22.

EPILOGUE
The big picture conclusion.

Let us sumarise the history of interconnected related key discoveries and concepts mentioned previously in both this book and volume one, which involved the greatest scientific debate between Albert Einstein and Neils Bohr.

The standard format for the acquisition of knowledge in science is empiricism: observation and experimental result—not only theory. However, this refers to observation from the physical body, consisting of physical senses and scientific instruments, and we should call it physical empiricism. Why is this necessary; what else is there?

This is precisely why Darwin posed the very profound question: How can man understand nature if man is part of nature? Darwin was referring to his own theory of evolution or any other scientific theory of the universe. He knew there was something missing but didn't know what. In fact, he was unintentionally questioning empiricism—not realising there was any other form of empiricism. Then quite remarkably quantum physics (now some eighty years ago) came up with not only the same query but eventually answered it.

We arrive at the conclusion that the total cosmos is a brilliant system of many universes of varying degrees of order for the exploration of the Absolute (Section 6), that is, the quantum realm of infinite possibilities. We have seen that the Absolute, which is beyond particles, waves, space and time, contains the non-quantifiables; the aliveness characteristic, sentience (basic consciousness), and the experiential property. However, every thing that comes into manifestation from this is structured from particles, waves, space and time. It is potentially quantifiable (can be detected, measured) and is a *relative* condition.

The system that is utilised for this cosmic feat is the fractal method, which note is inherent within the hologram, as explained earlier. The fractal system provides different degrees of order, forming a dimensionalised spectrum of increasing frequencies from the lowest order fractal (compare twig-level analogy) to the highest, the top of the 'pyramid' (compare tree trunk) and then the ultimate One or Absolute.

Note that the Absolute does not have specific beingness ('personality'/God); it is unconditional and neutral, beyond all quantitative description. Inherent within it is the infinite wisdom but in its primary state it is not formatted (recall analogy of ocean model in which manifestation is a product of formatting the water with wave patterns). However, it is logical that one might consider a self-analysis taking place in which self-observation creates objectivity and separateness. We then have a degree of finiteness from the Infinite and the emergence and formatting of personalisation, as described earlier in the book.

Thus as we construct manifestation from the Absolute, the exploration of this total subjectivity would logically lead to a qualification of a single self-aware beingness, which through objectifying many experiences from its infinite possibilities, starting with a small relative degree of objectification (at this very high level), a personalisation of these characteristics is formed. But note this degree of objectivity is chosen—the objective 'unconscious' side of the manifestation is not actual unconsciousness; the Being can look from any viewpoint. In other words, a personable God state would form, compatible with the religious expectation.

As we descend the fractal strata of increasing degrees of separation, objectivity and unconsciousness, going down the scale, creating fractal degrees of order, and increasing degrees of limitations (involving constructive boundaries), the purpose is to explore simultaneously a wide range of life-form behaviours, characteristics, and intelligence, along the infinite fractal scale, which has 'steps' or boundaries at ordered divisions, such as 3D, 4D, etc. These are natural boundaries, for instance, between the 3D waveband or spectrum and 4D. It is necessary to have an orientation of 4D with respect to 3D, otherwise the worlds of 4D would draw into 3D; actually merge.

Note that these natural *boundaries* (built-in designs) do not hinder the ascending inhabitants. In addition, *barriers* also exist that are unnatural and quite severe. These negative barriers, designed to arrest proper evolution or ascension of species up the fractal dimensions, are numerous thought-form creations from brainwashing the civilisation with existential limitations (for example, that there is nothing beyond 3D, no life after death, no higher consciousness and worlds), but also there is the presence of a holographic insert, surrounding the planet for thousands of years, referred to as the NET.[1] Fortunately this barrier is slowly being dismantled at this time by the so-called Guardian races.[2]

Thus the multiverse is structured, giving many levels of degrees of order (recall that degree of order is true basic intelligence of all creation) for the Absolute to explore its infinite possibilities by expressing itself in countless living forms.

Significant life-forms, such as humans, function on these different levels simultaneously. Why are there so many universe fractal levels? If there was only one level, there would be a mixture of infinite intelligence and low intelligence on the same worlds—impractical and inefficient. Organic life forms of sufficient growth (such as humans) are given free will. This means within the *natural* ordered boundaries and limitations of the fractal system (such as between 3D and 4D) non-ordered or random elements can be organised by, say, humans to creatively regenerate the higher frequencies and order (say, of the next level). This will create nonselfish qualitative states and develop corresponding frequency patterns of identity of individuals, which automatically select and ascend to the next level. In other words, the individual lives a life (and many other lives) and formats a unique individual personality based on behaviour and acquired knowledge that accretes data stored as frequency patterns (whether of DNA or consciousness).

This forms the individual ID—an immensely complex frequency spectrum/matrix—that determines the destination in the future life (or lives); governed by resonance and compatibility of the ID with the frequencies and degree of order of the next dimension. If the resulting ID spectra are non-harmonic, based on selfish or criminal behaviour, the individual will be drawn towards those corresponding dimensions—such as an environment of criminals. Also the environment would be less harmonic and beautiful, in other words, environmentally unfriendly.

Existing in the lower fractal orders, such as our 3D, it is inevitably more difficult and potentially hazardous. This follows from the fact that consciousness reduces from the Absolute of total singular consciousness via formatting stages down the hierarchy of dimensions of increasing unconsciousness.

Accompanying this is fragmentation, objectivity, and decreasing mind as the two sides, objectivity (the external world) and subjectivity (the inner self) continue to divide (see Figures 7 and 14). The environment becomes more independent of the mind and consciousness. Separation and objectivity are proportional to unconsciousness. Unconsciousness, such as matter, is simply another form of the Absolute made up of tiny units of consciousness, forming composite unity (parts not in phase). Even an atom is a small unit of the Absolute or consciousness but with a fixed template/blueprint 'telling' it to be only an atom—no freewill is given or complex bio-energetic systems generating a sufficient degree of order for self-determined awareness.

In this gigantic process, it is interesting to note that inherent within the procedures of manifestation of this creation are probabilities for eliciting the negative elements of life. For example, for a manifestation and an existence there must be stability; a degree of permanence—recall earlier reference to the sensitive balance between stability and instability (flexibility) of the wave patterns. Thus over focussing on permanence (caused by ego, insecurity, covert manipulation, implant programs, etc.) can bring about a stagnation within evolution, or more locally or individually, retardation of changes towards greater knowledge, growth and development.

We gave examples of this principle as it applies to expansion and consolidation of knowledge, and the learning pattern, in which emphasising specific learning can cause solidity and inflexibility of the learning pattern, preventing increases in ability. Overall, this means for ascension, the security of permanence and preventing change degenerates into unwanted persistence, or being mentally stuck. All these characteristics, seemingly aiding stability (only), will stifle the evolutionary process.

A further vital creation step is the mechanism of counteraction to create degrees of objectivity within the original subjectivity described in Section 13. It was mentioned that this action/-counteraction is also the anatomy of all problems. Thus the mechanism of the persistence of problems is an inherent trap, which is available when consciousness strays from alignment of the intrinsic harmony of cosmic physics. (Note the relevance of the expression in Christianity of 'not doing God's will'.)

Take a look at the hologram. We explained in Section 8 its property of the whole image or information being contained in every part. These parts are different degrees of wholeness or order. All portions from the smallest to the largest are in phase, each supporting the whole and every other part.

A cosmos structured on the basis of the holographic synchronistic fractal hierarchy would have harmonic templates and blueprint. This perfect energy system might be considered to have more integrity than we do. When matured life forms enter the universe system they would know the laws (physics/spiritual).

What are the consequences of neglecting to abide by these laws? Any responsible life unit/entity incarnating into an existence and, say, behaving harmfully to others automatically severs the inherent underlying interconnectivity between the two beings. For humans, this also cuts our energy fields with the planet, sun, galaxy, and universe. This violation of these harmonics is recorded within the universe and man's mind. If enough beings behaved this way, it would ultimately destroy the universe. Thus the cosmos has built-in immune or override systems to protect not only itself but other life.[3] We thus have karma, somewhat like an antivirus mechanism, which appears to function by phase-conjugation, well-known in physics. The negative event is recorded by the frequency pattern, which is radiated out to the universe, and the universe sends it back in reverse (time reversal or 180 degrees out of phase; or simply upside down). Mechanistically two identical super-imposed wave patterns, one upside down will cancel each other. However, since we are dealing with life, the consciousness must participate in this remedy. In fact, it must recognise its connection with what it has done. Education on mind matters is so ignorant (but controlled) on this planet that karma generally simply does not work (in particular, since the 'victim' often demonstrates blame, revenge, hostility). As a result, the negative return of the frequency pattern continues to repeat. If consciousness does participate, cancellation is successful and the incident will not repeat.[4]

In addition to this ascension status (the identity spectrum) that determines the level of qualification for future existence of the individual, there is also the question of what level of truth or degree of order can be detected from a lower dimensional status. As we have described, particularly in volume 2, a lower order observation cannot detect a higher order owing to 1) lower frequencies cannot detect high frequencies, and 2) where there are higher-order coherent structures, a lower degree will 'collapse' the wave; that is, quantum reduce the higher order to the lower. This is like not detecting the unity (of the coherent particles/waves) and only selecting (quantum reducing) the parts (and interpreting this as parts only stuck together by forces and not coherent). This presents another safety mechanism to prevent lower intelligence from acquiring a higher level, which would not be fully understood and possibly misused. Higher knowledge and its (mis)application is thus protected.

Thus the consequence rule on a long-term basis will act as a guideline and provide a qualitative selection of what level of consciousness of life forms will be able to ascend to the next level —thus providing a natural security system, which is self-regulated (and self-responsible) by the individual's behaviour.

One could conclude that the cosmos has the built-in ethic that for every action there is a consequence. When the action is destructive, then remedial and compensatory mechanisms are activated to protect the cosmos and other life forms within it.

Notes
1. NET stands for Nibiru Electrostatic Transduction field. For further information, see *Engaging the Extraterrestrials: Forbidden History of ET Events, Programmes and Agendas*. Also *Voyagers* vol. 2 by A. Deane.
2. Books: *The Original Great Pyramid and Future Science*, and *Voyagers* vol. 2.
3. Workshops from the Guardians by A. Deane.
4. www.nhbeyondduality.org.uk. Article: *The Mechanics of Karma*.

APPENDIX A

INTEGRATION, DIFFERENTIATION AND QUANTUM REGENERATION

True integration means perfect wholeness or unity, which is inherently holographic, and which in turn means greater power is available to the part.

There is an intrinsic relationship between integration and differentiation, that is, between the whole and the part, but it requires a special relationship amongst the parts or, in other words, a true whole. The key is quantum regeneration.

The greater the wholeness of a harmonic system, the more detail it can express in the part (this is always lower-dimensional). That is, the inherent wavelength is shorter (or frequency higher). The wavelength acts as a single whole, a quantum pulse; the smaller it is, the smaller the 'bit' size. When the parts are in a special relation, such as resonance, then integration is holographically related to differentiation. True unity—the ultimate would be the Absolute (Section 6)—is inherently holographic (potentially). Thus the whole has a higher frequency and can project a smaller 'bit' size in representing itself on a lower level of organisation (note that in this example we are not considering that the wholeness has particles with 3D frequencies).

In skills, the greater is the integration of information (bits) in the learning pattern, the greater is the control of the detailed movements (differentiation). In learning the piano it is sometimes expressed as developing independence of finger action, and therefore the greater the 'dependence' (unity) the greater the independence.

We may find that all learning involves resonance: the integration of parts to form greater wholes or quantum states of data. Academically we merely think of two items as associated; that is, stored side by side and connected in some way. This would be the case at the 3D level but underlying this within an inner-space layer there will probably always be a single state, representing the whole of these two parts. Stimulus-response mechanisms would also be included in this group of phenomena (the link between stimulus and response). It is interesting that quantum physics has long since applied this principle with the wave function but it is only considered to be a mathematical convenience or fiction.

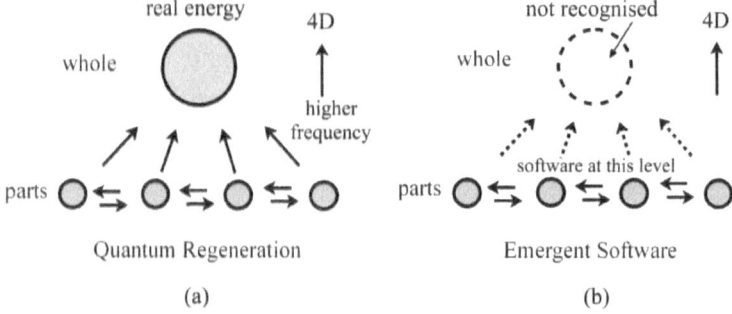

FIGURE 33: Comparing quantum regeneration with orthodox science's emergent software.

As already indicated in earlier sections the true whole is generated from the parts when the parts are harmonically related, such as by resonance of frequencies. Although science has the description 'emergent software' to express that the parts have generated a new condition, for instance, a new program, it doesn't recognise the higher-dimensional aspect and that a quantum regeneration of hardware has taken place. It only recognises the new program as software, nonphysical, instead of a new physical state of higher frequency. See Figure 33.

It may appear so far that we are compromising our understanding and partially supporting current materialistic science by introducing quantum regeneration, which apparently shows that creation is 'bottom-up' and not 'top-down' (such as from a god). Didn't we show that the parts create the whole? We did, quite definitely in the case of the laser (see Appendix B). This at least in principle has a major parallel with mainstream evolutionary theories. However, as already explained these higher states of wholeness already exist in what we, at the lower level, might call 'in a virtual state' (not yet made into a reality at our 3D level).

This quantum regeneration principle literally lifts energies above 3D and will do the same with consciousness and knowledge, that is, bypass the boundary between the 3D and 4D spectra, in effect lifting the 'lid', which has been made secure and a tightly-closed barrier by countless facets of 3D existence, in particular, limited belief systems, the scientific laws—even Christianity does it by encouraging the belief in an external God, contrary to what Jesus taught.

The bottom 3D level is quantitative if the parts are not in special relationship; that is, there is no quantum regeneration. The intellect and our current logic only handle the quantitative. Orthodox science is teaching us that everything can be understood from the parts—just as can a motor engine. The reason why the motor can be broken down into parts is because its parts—various components—are not in intrinsic relationship, such as frequency resonance and there is no true whole—just a composite, or simulated unity. When the quantitative state forms a composite unity (not a true whole), such as with the motor-car, the parts are held together by forces—nuts and bolts and welding. In this case the whole equals the sum of the parts, and the intellect's logic can handle this, since the whole can be broken down into parts and the parts and their relationship understood, with no limit placed on the understanding. None of nature works like this; the whole is greater than the sum of the parts, and in fact, already exists.

Science teaches that everything can be understood this way because everything is made up of parts. Even in emergent software this is only a non-physical program, and therefore a kind of illusion—this type of 3D wholeness does not connect up and activate higher states, it is still on the 3D level. Let's take a different example, one from the arts, for instance, in music. This is science's quantitative analysis of music.

An example in music, which we have partly covered in Section 16, would be when listening to a good piece of music. We only hear one sound at a time (even if it is a complex chord). There is clearly no music in an isolated sound (science is with us on this). There is no music in a sound 'now'. The mind holds on to immediate past sounds, anticipates future sounds and combines them as one whole, past, present and future over a short interval of time. Science has no argument with this. However, the process of perception of these sounds—past, present and future—creates, science says, the *illusion* of the tune/melody, the essence of the music. This is the 'emergent software'. But it is not an illusion. It is a real energy state: a frequency pattern; a quantum-regeneration. Science can't detect these high frequencies (we have covered this thoroughly in earlier sections). This energy state is what the composer created and then had to quantum-reduce to parts—the notes on the music score—as efficiently as possible. The listener then re-creates this aesthetic energy state by listening to the parts, then connecting together (not intellectually as science thinks) the past, present, and the future anticipated sounds, over a sufficient interval of time. This unconsciously then resonates with the whole essence of the music, assuming the music has sufficient integration/wholeness (a simple good tune has wholeness: a single quantum state of energy, a frequency pattern).

Also there is a similar principle in the visual arts. Consider a work of art, say, a painting of high merit (aesthetic art, not *other* forms of communication, *utilising* art). Now the bits of paint on the canvass are quantitative, physically, in that they are fragments stuck together by forces. However, the artist has placed the brush marks in a special order. Thus, the artist has created a qualitative effect from a quantitative state or medium. The appreciative viewer now looks at the physical surface of the painting; eyes may go slightly out of focus, the intellectual and conscious mind step aside and the 'unconscious' now correlates all the bits to quantum regenerate one whole state. Because the bits of paint (the 'quantitative') have been subtly ordered so as to be capable of correlation, then resonance can be achieved, resulting in a single whole energy that has (hopefully) complete unity and is a quantum state.

This can be thought of in the following manner: The artist encodes his idea in qualitative bits of paint ordered geometrically and by colour. It is then the task of the viewer to decode the energy pattern of the art and experience the 'meaning' by quantum regeneration of the parts to the whole.

This condition is of course in the mind. Thus a qualitative state in the mind of the observer is now superimposed on the quantitative state of the paint on the canvass—and appreciation has occurred. Of course we can't say much about the experiential aspects of this in terms of physics, except that it will be at least pleasing, will have a sense of unity, harmony and completion—nothing need be added or taken away.

We may note here that the integration and higher spectra aspects of good art and music involve a physics that runs parallel to true evolution itself, which is mainly expansion of wholeness and developing higher frequencies.

A very different example would be two people communicating and being 'on the same wavelength.' This is literal. Certain frequencies associated with the subject material of their conversation become entrained, that is, are the same and are in step. This quantum regenerates a single—even if intermittent—collective state that has frequencies higher and more energetic than the individual frequencies of each person, and greater than the sum of them. Two such people or a group working together in harmony ('on the same wavelength') can generate more power to be effective, than the sum of the individuals. This will be more than the square of the sum, as mentioned with the laser.

Collective effects of consciousness or species, and thought forms, would come into the quantum-regeneration category. The individuals generating the same frequency patterns, whether positive or negative, will quantum regenerate a powerful effect at the collective level, or even re-create the collective where this is weak and fragmented. This again is the emergent software effect, to which we have referred and which science doesn't recognise that there is a real quantum state of collective energy rekindled out of the virtual state hierarchy—not some abstract, non-physical, insubstantial software program.

The organism 'dictyostelium' is a good living example. These organisms periodically send out signals that bring together about 100,000 or so such organisms, which merge, forming a single organism, just visible to the naked eye. A new program is elicited and this entity then finds a suitable plant to climb where it extends itself as in a mushroom shape with most of its mass at the top and waits for a gust of wind to scatter the 'spores' into new foraging grounds. Note that the organism would already have an inherent collective, but so do humans. Also the connectivity between the individual dictyostelia, if over relatively significant distances, would be expected to be achieved by scalar (mind) waves rather than electromagnetic waves (or chemicals).

An excellent final example to present would be the quantum regeneration principle applied in advanced spacecraft propulsion. The craft's generator, which would be a scalar-field system, entrains all the atomic oscillations at the nuclei level of the body of the ship. That is, the frequencies are put in phase and made coherent (as with the laser, etc.). This instantly quantum regenerates a whole (collective) vortex oscillation in higher space (meaning of higher frequency outside the 3D range). (Note that a still more advanced civilisation would achieve this without a generator by using their minds.)

The atomic frequencies are now sub-harmonics of the single, higher and more powerful frequency. This powerful energy oscillation envelops the craft, and it is of the same physics as cosmic oscillations of natural bodies, such as planets, stars or galaxies, etc. Thus the craft can adjust and align its frequencies with those of cosmic bodies, such as planets, stars, star systems, the galaxy, etc. Depending on the phase relationship, it will be pulled towards or away from the cosmic body at very high velocity. In general, an out-of-phase relation would be maintained for intermediate speed control (see *The Emerging New Science*).

We see then that in life and nature there is a whole energy state that precedes the parts: the branch came before the twig and is higher in the fractal hierarchy. Thus there is not only the energy process of quantum regeneration, or bottom-up, but there is also quantum reduction, top-down. This actually means that the whole came first (which is obvious with the tree analogy, for example, the trunk), then fragmented to parts and the facility then arises for the parts to recreate the whole—but in the latter case, that whole, in terms of higher-dimensional energies, is potential. The dimensional framework, or template and raw spectra, is already there in a virtual state and constitutes the essence of evolution or ascension. This applies to the laser we described above.

The main point here is that ascension is a form of quantum regeneration in which parts (polarities in this case) come together to 'ascend' to greater degrees of unity. This is because a lower level of frequencies has activated a higher level and connected up to it. This is also a case of when the whole is greater than the sum of the parts. Thus particles and antiparticles coming together is one of the most basic processes of true evolution.

The ideal scientific example of quantum regeneration is the laser—Appendix B.

APPENDIX B

THE LASER AND QUANTUM REGENERATION

Can science really explain where the enormous power originates in the laser?

We gave a brief description of the laser property of quantum regeneration in Section 13; we shall recap here and present the diagram. We begin with a quantity or number of rays of white light. Quantity means the parts are separate, unrelated; the rays of light are higgledy-piggledy, are out of phase, that is, random. We now put the parts into special relationship; we put the rays of light into coherence so that their oscillations are in step in space and time—this gives the laser beam.

This is resonance and we now have quantum regeneration—a whole quantum state. The rays of light, although retaining their individuality and separate locations at the 3D level, quantum regenerate a higher-frequency oscillation (this is the vortex) and the so-called 'emergent software' appears as a real energy state of greater power (Q2) than the sum of the separate (Q1) rays (we know from physics that the higher the frequency the more powerful is the energy).

In Figure 34 we give a simplified example of five rays of ordinary white light. The rays normally will be out of phase, that is, random—not aiding one another. The calculation shows the total energy without quantum regeneration, that is, the whole equals the sum of the parts, compared with the total energy after quantum regeneration, that is, the whole is greater than the sum of the parts.

The New Science

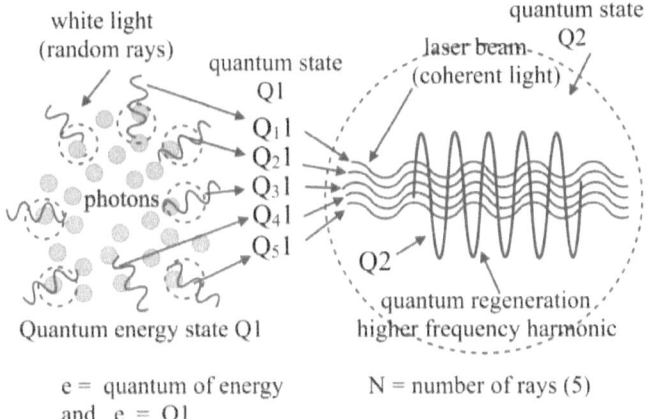

e = quantum of energy
and e = Q1

N = number of rays (5)

ORTHODOX SCIENCE
Total energy E of laser beam (of N rays) equal to sum of parts,
$$E = Q1 \times N = Q_1 + Q_2 + Q_3 + Q_4 + Q_5 = N.e$$
where N = 5
i.e.
$$\boxed{E = 5e}$$ (the whole equals the sum of the parts)

NEW SCIENCE
Total energy E of laser beam is greater than the sum (N) of the parts,
$$E = N^2 e \,[1 + \tfrac{1}{2} + \tfrac{1}{3} + \ldots 1/N]\quad \text{(square of N times a holographic factor)}$$
or
$$E = 25e\,[1 + \tfrac{1}{2} + \tfrac{1}{3} + \tfrac{1}{4} + \tfrac{1}{5}]$$

$$\boxed{E = 57e}$$ (the whole is greater than the sum of the parts)

FIGURE 34: New Science evaluation of laser energy taking into account quantum regeneration, compared with orthodox calculation.

Thus the total energy will be five squared, plus further holographic terms. This system is explained in full in the book *The Great Pyramid and Future Science* by the author.

This huge amplification applies for all groups in resonance in this way.

APPENDIX C

VISUALISING THE 4D VORTEX

Science never took the vortex theory beyond 3D and so missed out on the perfect supersymmetry that arises with this higher-dimensional view.

The diagram in Figure 35 shows a method of visualising the conversion from the 3D vortex to 4D. We see in Figure 35(a) the typical 3D double-vortex polarity. For simplification purposes visualise it as two cones, one inverted relative to the other, shown in Figure 35(b). Now go to Figure 35(c). Remove cone B temporarily and expand cone A until the two radii of the dotted circle meet at the bottom and disappear.

One can initially visualise this now as a 2D circle on the paper but remember it is supposed to be a sphere. See Figure 35(d). Now bring in the vortex spiral and try to visualise it as shown in Figure 35(d) but further that this is really a sphere and the vortex fills the sphere as it spirals all round into the centre. This is only one pole.

So far we might see that this spiralling-in action has no where to go; there is just an imaginary point in the centre of the circle (sphere). What about the bottom half of the double vortex spiral? What's happened to that? Again go through the procedure using the B cone as shown in Figure 35(e) as we did with cone A. We finish up with Figure 35(f); note the direction of the spiral, which is opposite the first. We must now imagine that the Figure 35(f) sphere is beyond and through the centre point (in the anti-universe side) in Figure 35(d).

From the viewpoint of A, sphere B will be within it, but from B's viewpoint, A is inside B. The direction through the centre of A to where B is, is a 4D direction, at right angles to each of our three dimensions. All natural entities, atoms, planets, stars, universes, must have this configuration; that is, both sides, in order to exist as a stable form.

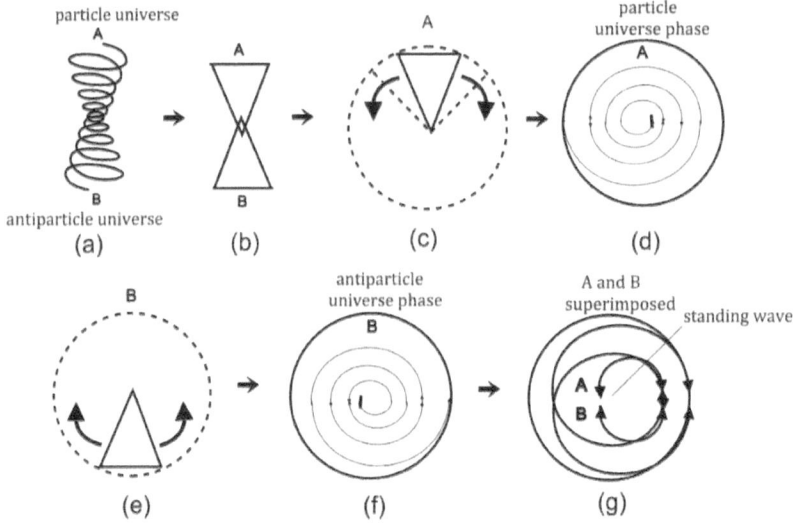

FIGURE 35: Visualising the 4D vortex starting with the 3D vortex.

The vortex for our planet will give rise to gravity; the sphere (Figure 35(d)) pressuring inwards to the centre of the planet. Clearly we only see one pole to gravity; there must be two poles. The other is on the B side in the anti-universe.

Figure 36 shows a secondary vortex induced by the first; that is, the lines of force encompassing the spiral coil. One can see from the diagrams that the appearance of the doughnut shape begins to emerge—the lines of force delineate the doughnut shape. However, this is still only a 3D (doughnut) shape. One must use the steps again in Figure 35 to visualise the 4D doughnut. This is much more difficult though than the first visualisation in Figure 35. Just imagine that the 4D doughnut would be something like a thick spherical shell around a solid, much smaller sphere (inside), with a space between the two—which is a shell of empty space; this is the hole in the doughnut.

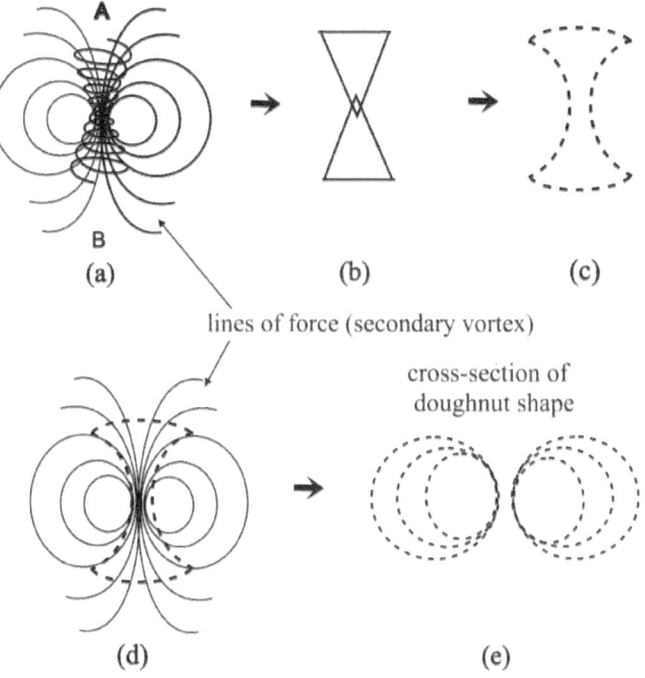

FIGURE 36: Visualising the doughnut shape from the 3D vortex.

This isn't all there is though, for example, if you moved inside the smaller sphere and through its centre to the anti-side and looked back, the small solid sphere would now become the shell and the shell now the sphere inside it. Thus we appear to have two spheres, one on our side and the other in the anti-side. However, as one moves outwards from either of the two surfaces the energy connects into 4D (that is, further out in space) and they both join up—become the same.

APPENDIX D

QUANTUM REDUCTION AT DIFFERENT ORDERS OF THE FRACTAL SCALE

The degree of order of the hierarchical fractal levels of the multiverse system gives us a metaphorical ladder to ascend one step at a time and encourage the individual to develop spiritually and qualify for the next higher level.

The fractal levels provide the major fixed divisions in this hierarchy of orders. The divisions, however, have large gaps in them (such as 3D to 4D); jumps in the gradient of order. Within these levels we have randomness and many possibilities, and the background fractal-matrix gradient.

The evolving individual, however, provides the gradient—fills in the gap, so to speak. Consider an analogy: the ground-floor worker gradually acquires the qualification bit by bit to achieve the jump from the ground floor to, say, manager level. In life, this is the accretion of frequencies from the background, unified quantum field, plus the ordering and integration of these frequencies and frequency patterns into the individual's personality, leading to the required qualification to achieve the quantum leap to the next level.

Thus the personality has his or her own frequency-spectrum identity and by means of the physics of wave patterns this energy structure, consisting of mind and consciousness, is drawn up or down the dimensional structure. Upwards requires greater

The New Science

coherence, downwards fragmentation and decoherence—the latter automatically drops in frequencies.

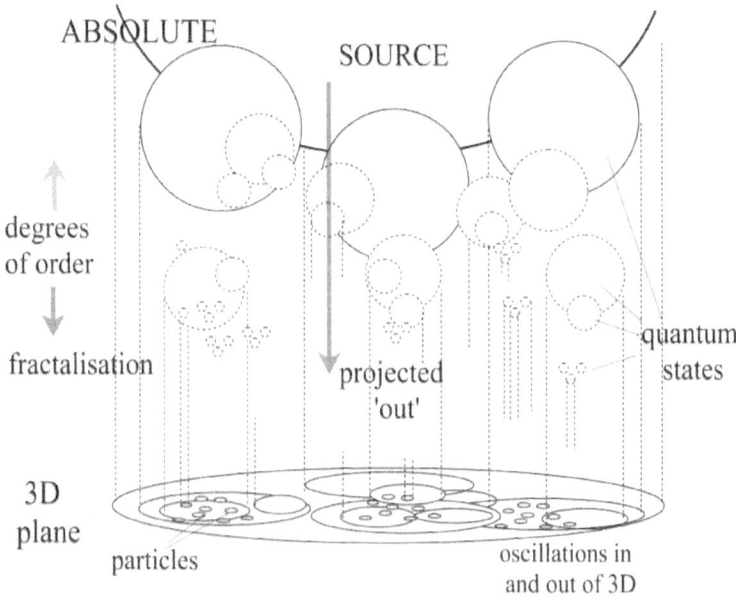

FIGURE 37: The spot-light analogy in the 3D plane. Spheres within spheres, all within Source.

Amongst the randomness and disorder between fractal divisions, but more so at the lower levels, there are thus fractal orders within orders (recall that the basic hardware is always spheres within spheres—higher dimensional vortices). In Figure 37, quantum reduction takes place from the larger spheres to the smaller ones. In Figure 37 we see that the entity or device interacts with a particular level of the universe and receives back from the universe, information governed by the context of the observer. It is all about orders interacting with orders and the subsequent quantum reduction of higher orders. These all take place within the primary Copenhagen collapse of the wave function. This is the probability selection of first order which manifests the material backdrop. Secondary collapses then could occur as was illustrated in volume 1 (Figure 25, not shown here). The diagram wasn't easy to follow, in particular, it relied to some degree on colour in the original e-book which significantly aided the understanding. Nevertheless we have a couple more diagrams here for anyone wishing to pursue an understanding of this complex interaction.

Figure 37 has the spheres within spheres structure, though here they are projected out for clarity, but the fractal scale, dimensions and frequency spectra are inherent.

Figure 38 attempts to show that there is not only the basic quantum reduction (the Copenhagen interpretation) in which the interaction of the observer with the quantum realm selects the appropriate probability, which (for us) manifests as our 3D reality, but that this quantum reduction from a high order to a low one occurs all the way down the fractal scale. A lower order impinging on a higher automatically selects its own level inherent within the higher order.

If this was all, then one would ask, How can one evolve (to a higher order)? If the observer is an artificial machine, the answer is, it can't. A human, however, can quantum regenerate elements from the environment. Note that relative to consciousness, the mind structure itself is, in effect, a mentally-empirical environment. By putting order into parts (for example, resonance) a greater (carrier) wave function is quantum regenerated, corresponding to a higher position in the scale. But where does this higher order come from if the observer resonates with and draws in its own reality level? Notice that it is not 'generate', but 're-generate'.

Thus the human's extended mind, of holographic and fractally-organised particles and waves, can always go one higher than its immediately-observed universe environment. This means people can put higher orders into lower orders of their immediate environment as they quantum regenerate to the higher order, drawing their mind/consciousness up to the next higher level by coherence and resonance.

The New Science

Thus the particle/antiparticle polarity in 3D is within, or carried by, the polarity of the next level, and so on, that is, polarities within polarities up the hierarchy, until 100% subjectivity is reached of no separation and no polarity.

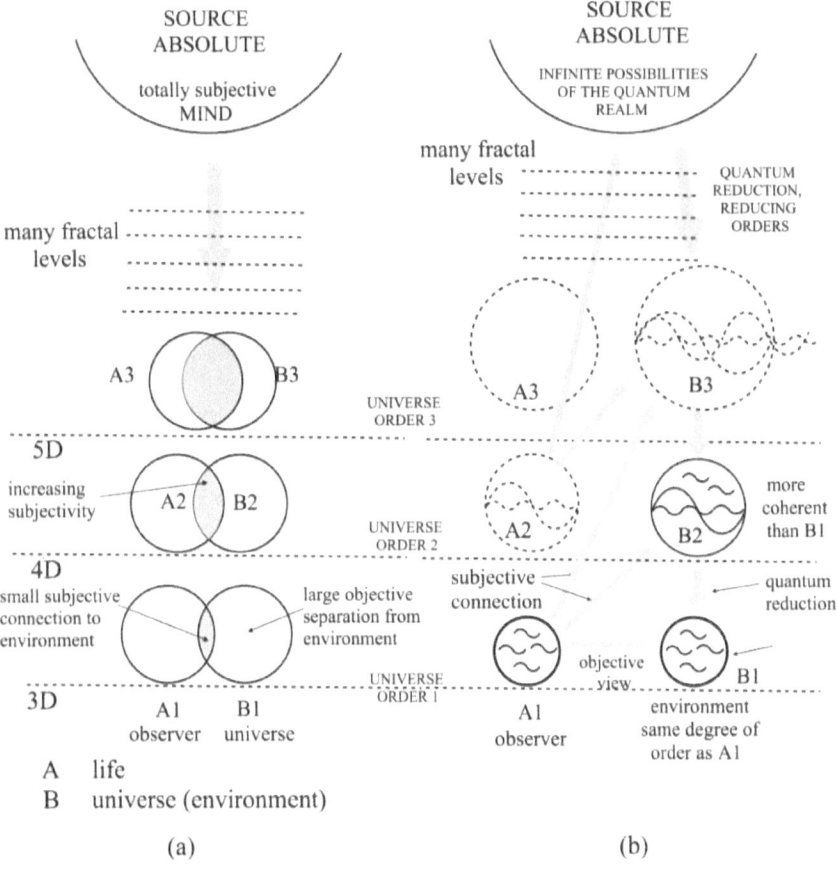

FIGURE 38: Universal fractal orders, showing the true nature of evolution/ascension with the changing relationship between the observer and observed, towards a complete resolution of the duality in a single 'mind' at the top.

APPENDIX E
SCALAR ENERGIES

Scalar physics research is not encouraged in universities for the similar reason as to why antigravity isn't taken seriously. Moreover, the latter would lead to expulsion of any professor engaged in such research.

Academics are still sceptical about scalar waves; the subject is not easy to authenticate, that is, to prove their validity experimentally. However, as a starting point it ought to be obvious that if there are 3D waves, such as electromagnetic transverse waves, there should be waves (though less detectable here in 3D) that oscillate in 4D. This is simply higher-dimensional geometry—not taught so that people, in particular, academics, won't grow too smart (recall that the answers to everything lie in the higher dimensions and not in this 3D).

It may be helpful to the reader to begin by recalling the ocean analogy. The sea (or water) represents all of space and in our context here, it represents the aether. Furthermore, let's not forget what quantum physics is telling us, that all bodies (all manifestation) are like shapes and variations in the structure of space (but the aether was dropped from science, remember?).

The continuous energetic activity from the infinite interferences from the waves of universal bodies (planets, stars, galaxies, etc.) generate the virtual particles, mini-black and white holes and the potential energy variations. This is underlying all observable particles (such as photons and electrons) and everything we call *mass* (recall: even electromagnetic waves use mass particles (electrons)). Thus all this basic activity is scalar. It does not consist of force-fields, since these require the presence of mass particles. As with the gravitational field, which is scalar, there is no force until a body is present in the field that is being resisted or impelled (remember, in free fall there is also no Newtonian force). Similar for particles in the aether—if they go with the flow, governed by potential gradients there is no Newtonian force. Thus scalar activity is ongoing but can be engineered (organised flows to be tapped).

Now we have previously mentioned three types of waves: transverse electromagnetic waves, longitudinal waves, and gravity waves. The transverse waves, such as light waves, oscillate at right angles to the direction of motion of the wave. Longitudinal waves oscillate along the forward direction of the wave (as in sound waves). The gravity wave oscillates at right angles to 3D itself. This latter mode of the wave is what is called a scalar wave. The first two have charged particles (for example, electrons) in oscillatory motion in the aether, but the third (scalar component) does not basically involve these observable particles—this scalar wave oscillates the aether medium itself.

It is basically a 4D potential energy wave, which means it oscillates in time. It is manipulating the basic aether from which all forms are created (recall the ocean analogy again). Thus scalar waves are associated with quantum mechanical fluctuations of the aether, and also the zero-point energy state. Ordinary electromagnetic wave disturbances will be accompanied by scalar potential energy activity, but with the presence of particles; the electromagnetic wave is detected as propagating transversely in the 3D plane. How is a scalar wave generated?

We have previously mentioned a similar property of what happens when a particle and antiparticle combine; that is, two opposite particles in phase but mirror image. They do not naturally annihilate one another with a sudden release of energy (a matter and antimatter explosion as science teaches). This may occur under unnatural and artificial conditions, such as in the laboratory or in a mutated environment. However, under natural conditions the pair unites to form a higher-frequency particle, which immediately appears in the next fractal level (higher spectrum), where it divides again into particle and antiparticle according to the higher-order frequency codes.[1]

Thus the salient feature here is that two opposites may not cancel out (completely). Similarly, if we have two equal (same frequency) and opposite electromagnetic waves, they will cancel, but we are left with the mysterious scalar wave. (Originally mainly researched and developed by Tesla, but today is surrounded by secrecy since it has been utilised for scalar weaponry.)

Scalar energy is obviously basic to electromagnetic energy since the former is a property of the aetheric potential energy gradients from which originally the particles in electromagnetic activity were produced (Section 19). Scalar energy is inherent in the vacuum potential of all space. As stated above and in the section on gravity, the gravitational field is a scalar field and the gravity waves are 4D, which are oscillating at right angles to 3D. When science measures the gravitational force, such as weighing an object (which is stationary relative to Earth), or when measuring its momentum and inertia, the scientific method interferes with the body—breaks its 'free fall' of coherence with space in which its motion is governed by the scalar field. Thus as we mentioned earlier, the gravitational field is not basically a force-field; the force only enters into the equation when the measurement system is applied. Similarly the force fields of the electromagnetic waves are due to the presence of charged (mass) particles, which then hide the underlying non-force field scalar (no particles). In general, scientific measurements deal with the force aspects of the electromagnetic components: electric field and magnetic field, and these fields are measured in the presence of particles, such as electrons, which generate the force. Thus the Newtonian force is due to the presence of particles; it is not present in the underlying scalar field.

As an additional revelation, a particle allowed to move freely in the scalar field, that is, not doing work, is coherent with the field, just as is the body in the gravitational field, and thus in this case (corresponding to the object in free fall, that is, coherent with space), it is driven by the scalar potential gradient—that is, goes with the flow; which is not a force.

Note the Russians tend to use the word torsion (rather than scalar); a descriptive word, though tends to have the context, 'force' since the scalar potential is governed by stresses, compressions and expansions of the aether. Other names are quantum fields, gravitational fields, Newtonian and tachyon fields, Tesla and Maxwellian (Teslawellian) waves or non-Hertzian waves, and longitudinal waves.[2]

Large amounts of energy can be transmitted over considerable distances by first creating scalars from the electromagnetic fields, and then restoring the electromagnetic energy from the scalar at the destination. A pair of electromagnetic waves phase-conjugated creates/releases the scalar. Then two scalars, carrying information but not carrying the electromagnetic energy, without significant attenuation are sent to the target. At the target, the scalars of different potential couple to produce any amount of specified electromagnetic energy (for example, light and heat).

In Section 21 it was mentioned that as per current science's light velocity and its constancy, it doesn't follow that we would know the distances away of sources of light, such as far-distant stars, since as we have indicated, the light velocity is not only variable but will have underlying it a fractal scale of increasing speeds of transmissions (as we go from the electromagnetic transverse wave to the scalar wave (3D to 4D range) of ultimately, instantaneousness). Let's take a more local example, which also demonstrates the scalar properties and conversion to electromagnetic transverse waves.

On the basis of the scientifically-accepted light velocity we calculate that light takes about 8½ minutes to travel from the Sun to Earth. But we only measure it at its destination where the light enters the scientific set-up, or see it at the human eye. The Sun will emit higher-frequency components forming a standing wave from the Sun to planets or their atmospheres. There will thus be a scalar standing-wave component which is oscillating (in time), for instance, in our atmosphere, instantaneously. These scalar waves come into 3D instantly (from 4D inner space) and interact with the atmosphere and quantum-reduce from 4D to 3D, creating electromagnetic transverse waves of heat and light. The interaction occurs in the atmosphere, causing electromagnetic ripples which spread and propagate. This will, of course, occur in front of the eye or detecting instrument, where it will be measured at the standard velocity of light (hence the misleading deduction that the light took 8½ minutes).

Note that this is an example of transduction from scalar to electromagnetic activity of a constructive nature. As mentioned previously the interacting components can carry huge differences of potential and be used as weapons, causing massive explosions (of heat and light).

Now electromagnetic waves of longer wavelength become more longitudinal. The transverse compression and rarefaction oscillation becomes more longitudinal as we consider the propagation of longer wavelengths. The compressions become more in the forward direction; this opens up the possibility of transduction of longitudinal waves to give scalar longitudinal waves. It would seem logical that by lengthening the transverse wavelength (for fixed velocity of light), the transverse direction of compression orients more to the front.

A rough guide to the transverse and longitudinal activity is given in Figure 39. One might use the analogy of swinging a rope attached to a fixed point. The two nodes could be points on opposite sides of the Earth, or transmitter/receiver antennas. The standing wave A shows the scalar oscillation to be only in time (vertically), but would be virtually instantaneous. The slower scalar wave B

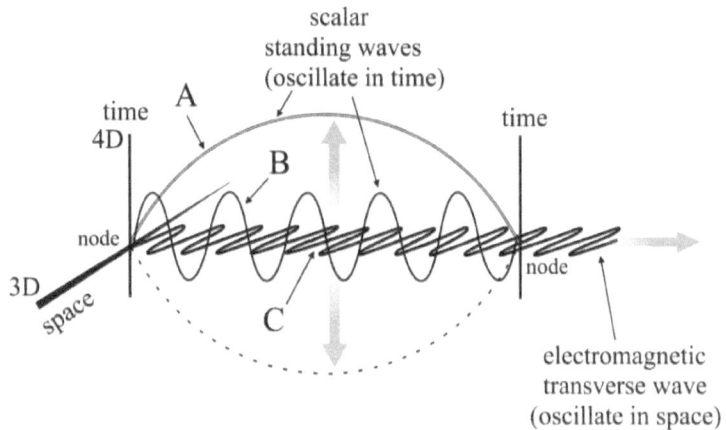

FIGURE 39

demonstrates both transverse and longitudinal components. Wave C shows the 3D transverse characteristics of the ordinary electromagnetic wave.

Thus the scalar is longitudinal but waves in 4D and time. This means the medium itself is oscillating (in time). It doesn't basically have spatial frequency as does the electromagnetic wave. The electromagnetic wave is transverse, which mean the *particles* are oscillating transversely, as opposed to the body of the aether moving in longitudinal waves along its direction for the scalar wave.

This is why the scalar is so powerful. All creation consists of frequency patterns, and basically each pattern has a sound and a light component (it is a shape in the structure of space). The sound essentially moulds the form of a body, and the sound is the scalar component. Scalar sonic weapons are the ultimate (our technology hasn't quite got there yet, thank goodness).

Thus in scalar activity the *aether medium itself moves* by compression and rarefaction (not just charged particles doing work). This is key to the understanding of scalars.

We may see that it is not surprising there is resistance from some academics to the scalar information, since the aether notion was dropped and anything to do with abundant (or, in fact, perpetual) energy has been suppressed. Electromagnetic energy and theory is accepted since it attributes the behaviour of the system to charged particles, totally ignoring the aetheric scalar potential gradient carrying the particles (and creating them as in virtual particles, or has already created them, such as with an electron). The field of an electron or, say, an ion (charged atom), is a potential scalar energy source, if one could tap into that flow, and use it, it would be perpetual. But our systems will cancel the charge with an opposite potential and the scalar source that the artificial system created, such as batteries (which then need recharging), is cancelled.

The 4D component of scalar energy enables it to act on the nucleus of atoms; whereas electromagnetic energy only acts on the peripheral atomic electrons. This means scalar energy can cancel inertia and open up new vistas of propulsion systems.[3] It is also correspondingly potent in effecting biological systems, for benefit or weaponry. Further properties are that scalar waves can pass through anything, can imprint their information into space, and reduce objects to dust. (In the '9/11' disaster, what caused dustification of solid materials, 'melting' of steel girders without heat, levitation effects of vehicles and citizens, etc.?)[4]

In summary, the basic character of the aether is a disorder of infinite interacting waves from significant bodies (such as celestial) creating patterns of compression and rarefaction (potentials). This then generates intersection points of potential, that is, nodal points of reinforced or cancelled waves, creating regions of higher and lower aether density, which then create fleeting mini-black and white holes and virtual particles in general that interact with and carry particles, electrons, protons (formed from more reinforced wave intersections; see Section 19). In an atom the electron and proton are prevented from pulling together, which leaves (and creates) the electric field. These particles and holes are all vortices; the small ones tend to be offshoots from the larger ones (Figure 4).

Only potential gradients exist in the 'vacuum'/aether.[5] Electromagnetic waves basically are not force-fields but potential gradient fluctuations. Scalar waves are not limited to the velocity of light. Waves are in the basic aether (4D) with no particle contributions, that is, electrostatic charge *is* the potential differences (potential meaning differences in compression and rarefaction).

Tesla had the correct understanding of radio transmission but it is not even recognised today (since it relates to the aether, which in turn relates to scalar that forms the basis of the forbidden free-energy field).

The radio transmitter emits electromagnetic transverse waves from the electrons oscillating in the antenna. However, electromagnetic waves attenuate and lack practical application. What happens is that some of these longer waves become longitudinal, bringing in (revealing) the 4D scalar component, which has little or no attenuation. Nevertheless these scalar electromagnetic waves will also be accompanied by some electron activity, producing ordinary electromagnetic transverse waves. Now at the receiving end, scientists detect only the electromagnetic waves, and the scalar remains hidden.

Quantum physics had it right with the requirement of an aether medium, but then relativity intruded and caused quantum physics to contradict itself regarding the definition of mass, energy and force. Electrons and protons, etc., are only shapes (wave and vortex patterns) in the aether and move along potential gradients. Thus we see that the scalar energy is fundamental; its 4D component links up with the whole inner-space dimensional fractal hierarchy of perpetual energy—one of the greatest crimes committed on the human race is educating the masses to believe that there is no such thing as 'perpetual'.

Notes

1. Unity of particle and antiparticle was described in Section 3 (page 31).

2. Book: *Tuning the Diamonds: Electromagnetics and Spiritual Evolution* by S.J. Rennison.
3. Book: *The Emerging New Science* vol. 1.
4. Book: *Engaging the Extraterrestrials: Forbidden History of ET Events, Programmes, and Agendas*. www.nhbeyondduality.org.uk.
5. *Toward a New Electromagnetics, Part 4: Vectors and Mechanisms Clarified* by T. E. Bearden. Tom Bearden, much maligned by some of the members of the academic community, is a protagonist of scalar physics; if we lived in a saner world, he would have been nominated for the Nobel Prize.

APPENDIX F

ORTHODOX AND NEW SCIENCE COMPARED
(Repeat from volume 1)

ORTHODOX, MAINSTREAM SCIENCE	THE NEW SCIENCE
Non-harmonic science and technologies.	Harmonic science and technologies.
Three-dimensional universe.	Multidimensional universes.
Evolution from a random condition and progressing by natural selection, and applying only to organic existence.	Evolution from ordered seeding imprints for all life and natural bodies through the dimensional structures of the multiverse.
Not handle synchronistic phenomena.	Existence is inherently a synchronistic system.
Linearity dominant.	Nonlinearity and inner space underlying all creation.
Holographic, fractal aspect of universe unrecognised.	Holographic, fractal universes recognised.
Chance, accidental events and luck.	Synchronous events and 'luck'.
Based on the quantitative and separateness factor with only simulated unity; not truly qualitative.	Based also on true unity and the qualitative, and its relation with the quantitative; recognises true unity. Recognises the soul as the next higher aspect (fractal level) of consciousness.

Three-dimensional logic (e.g., whole equals sum of the parts).	Multidimensional logic (e.g., whole greater than the sum of the parts).
Focuses on nonresonant energies.	Focuses on resonant energies.
Recognises emergent software.	Recognises quantum regeneration.
Validates objectivity and intellect.	Includes subjectivity and intuition.
Deals with 'surface' of universe, effects, illusions and linearity.	Embraces the inner structure of the universe and recognises the true nature of nonlinearity.
No science of mind.	Science of mind.
Considers consciousness and mind as by-products of the brain.	Recognises consciousness as primary and the brain (and mind) as a by-product of consciousness.
No life after death of the physical body.	Life after death of the physical body.
Computer system is dimensionally linear (algebraic encoding).	Computer system is dimensionally nonlinear (geometric encoding).
Not recognise relative zero and that the universe functions on a geometry of twelve.	Recognises relative zero and the intrinsic geometry of twelve.
Science laws considered immutable.	Science laws are relative to a limited context and can be bypassed.

The universe likened to a machine made up of particles stuck together by forces—see Figure 1(a), volume 1.	Universe of interplaying quantum states (wholenesses), in which not only particles but all natural entities have wholeness—see Figure 1(b), volume 1.
Universe is a closed system of Energy.	Multiverse is perpetual and open; is continuously created.
Static electricity is a closed system.	'Static' electricity is basically dynamic and open.
Learning patterns: spatio-temporal brain patterns and neural networks.	Learning patterns: 4D holographic templates that house programmes and convert nonlinear information into linear information.
Inertia and mass: innate properties; space and time are relative to one another (relativity).	Inertia and mass: contextual; governed by method of interaction; space and time are relative to the many other space-times.
Newton's laws: universal	Newton's laws: limited and can be bypassed.
Forces	Quantum Action
Gravity and the electric field are force-fields.	Gravity and electric field are not force-fields but scalar.
Theories promote ego, competition, and survival of the fittest.	Theory reveals cooperation, support and integration (for proper evolution).

Higher reasoning mind: intellect (no intuition).	Higher reasoning mind: should be a balance between intellect and intuition in complementary relationship.
Communication occurs through 3D space only.	Communication of life and the universe also occurs more fundamentally at a vibrational level (of frequency patterns).
No science of, or a proper recognition of, the species collective.	All species, races, have collectives; a synchronistic civilisation is possible.
Velocity of light limitation.	No velocity limitation.
'Father of Electricity' (false): Edison (rigorously campaigned against AC mains distribution system).	'Father of Electricity' (true): Nikola Tesla (invented AC mains distribution, plus modern electric light, radio (not Marconi), electric motor, turbine, and much more).
Teaches evolution as normal (not recognising presence of a non-harmonic parasitic evolution).	Recognises harmonic civilisations.
Artificial Intelligence. The big (incorrect) question: At what point in the advancing computer/robot system can we *consider* that it is alive?	Artificial Intelligence. The big (corrected) question: At what point can we say that it *is* alive?
Negative resistance, antigravity, perpetual energy sources, cannot be accommodated and understood with the limitation imposed by official scientific laws.	These 'heretical' concepts are a natural consequence of a harmonic science, such as is the New Science.

The scientific observer unknowingly quantum reduces coherent higher orders down to the material everyday level of organisation and randomness.	The New Science recognises the higher orders with appropriate levels of consciousness, *avoiding* the secondary collapses of the wave function (see earlier sections or volume 1).

APPENDIX G

KNOWLEDGE STRUCTURES

There can be considered to be qualities of knowledge. The higher the quality, the higher the degree of order and commensurate relative degree of truth; hence the presence of ubiquitous fractal systems.

The basic context for the foundation of knowledge must be the observer/observed relationship. In fact, as we shall see, or have seen, as indicated by quantum physics, there is no knowledge without the observer. In other words, there is no observed without the observer. Plato's statement that 'all knowledge is inborn' is thus not so strange; but it is then not 'intellectual' and quantitative. The latter is acquired through the objective relationship with the observed/environment/universe.

The definition of knowledge is an ongoing debate amongst philosophers in the study of epistemology. Webster's Dictionary defines it as 'the fact or condition of knowing something with familiarity gained through experience or association'. Experience is the basis of the method of empiricism,* and association is an intellectual and logical connection between ideas and sense data or memories, correlated with the external world; such structures then reflect or describe/explain reality. This latter approach was central to Aristotle's doctrine of mind rationalism. [* Empiricism: all concepts are derived from experience as opposed to rationalism, based on logical thinking.]

The steps to the acquisition of knowledge tell us something about the nature of that knowledge; even determines the knowledge —it doesn't already exist intellectually. It is a type of abstraction, since the universe functions on geometric intelligence, and geometric knowledge exists everywhere in 'tangible' form. Moreover, as we have indicated, direct and subjective perception can access this; a process that was more in line with the thinking of the ancient Greeks, who entertained a non-empirical approach and one we would now refer to as metaphysical and speculative that science today does not accept. Let us present a brief historical summary of these developments.

Following early Greek understanding of knowledge, which was mainly thought and inspiration, Aristotle was probably the first to bring recognition to the reasoning process as a necessary means to acquire knowledge. His logical approach was to define an object, construct a proposition about it, and then test the proposition by an act of reasoning. His laws of logic, which were of a deductive* nature, have provided a guideline for scientific methodology ever since. [* Deductive: to evaluate from existing established data.]

From this linear mode of logical thinking, reductionism* was a natural outcome, and the concept of chains of elements linked by association, from which has evolved associationism, led to the modern and prevailing psychology and philosophy. Thus, in general, the empirical approach to gaining knowledge has its basis in associationism and reductionism. [* Reductionism proposes that understanding is in the direction of smaller parts and supports the analytical method to arrive at truth.]

Further developments in logic were achieved by Roger Bacon who recognised the process of induction (in which the general is inferred from the particular). He proposed the need for validation of truth by sensory agreement amongst observers, aided by mathematics. Modern empirical scientific methods utilise both deductive and inductive processes of reasoning. Paradoxically, the downfall of Aristotle's metaphysics was due to his own extreme rationalism, which, along with Bacon's contribution, lost favour with modern thinking that emphasises empiricism and associationism with stronger materialistic connotations (but totally ignoring or not understanding the growing influence of quantum physics concepts).

In considerable contrast to Aristotle's associationism, Plato taught—recall the introductory comment—that all knowledge is inborn and its apparent acquisition by experience was an illusion—compare quantum physics which concluded that all objectivity was an illusion. Thus nativism* is closely akin to this philosophy. Other models of knowledge, which have sprung from a similar vein of thought, as contained within Plato's rationalism, are holism* and vitalism*. However, today with the establishment of empiricism and objectified experience, and a mode of thinking based on associationism, these philosophies are not given serious scientific inquiry, though nevertheless they do underlie or contribute principles to existing sciences. But if modern science exercised more awareness of quantum physics and gave it the recognition it deserves, notably the observer/observed relationship, even Plato's seeming extreme metaphysics would be seen in a different light.

[* Nativism is a psychology theory that states faculties, such as perception, are innate and not developed experientially—the New Science shows that reality involves both of these.

Holism emphasises wholeness and is non-analytical, for example, the whole is greater than the sum of the parts—compatible with the New Science.

Vitalism states that living processes do not arise from physical and chemical systems but are animated by a separate vital principle (New Science compatible).]

The traditional description of knowledge was the long-held account of knowledge in which three conditions must be met: 'justification, truth, and belief' (JTB account). This has since been repeatedly challenged (known as the Gettier problem/objections) as incomplete as it didn't cover all aspects of knowledge. The JTB account was originally attributed to Plato but following this there have been accounts in which he argued against it.

Thus modern thinking tends to evaluate knowledge as intellectually objective and to be arrived at via empiricism (through experience) based on observation or physical interaction. However, we shall see that taking into account higher evolutions as outlined in the New Science the *experiential* may become the ultimate knowingness. The experiential or subjective is direct and therefore must qualify as knowledge or knowing something—it can't be wrong since it is an experience, unless it is interpreted intellectually and then it becomes objective knowledge that may be true or untrue. Note that when we use the word 'experiential' here, it can refer to empiricism via objectivity or to direct intuitive knowing of which the latter becomes more obvious with spiritual advancement of the civilisation.

Nevertheless we have two basic categories: subjective and objective. And with a proper understanding of the observer/-observed relationship ('I/not-I'), as clarified by quantum physics, the objective or 'not-I', although treated as a separate informational structure, is simply another aspect of the observer 'I' (but the 'not-I' is unconscious mind). In the final analysis the subjective is the source and all objectivity is an illusion (not meaning that it is not real but that it is relative). We also see that knowledge cannot exist independently of the receiver, observer. The observer/observed relationship determines the ratio of subjectivity to objectivity which in turn determines the level of evolution of a race.

Thus even the objectivity of knowledge is embraced by quantum physics; it is one of the products of the 'I/not-I' relation of observer/observed. In summary we have:

1) Experiential/direct/subjective knowing—the more advanced a civilisation, the more this subjectivity or knowing takes over (replaces) the objective, avoiding the limitations of objectivity and experimental dependency.

2) Objective/intellectual knowledge varies with technology—detecting more phenomena through instruments and thus depends on experimental procedures. Ideally there should be a balance between (1) and (2). Ultimately (1) should prevail—no dependency, no indirect evaluations on the 'I' side; consciousness duplicates subjectively the 'not-I' side (turns the unconscious into the conscious).
3) Finally, depending on the evolutionary position the race is in (the fractal hierarchy), the directly perceived subjectivity is translated into objectivity, such as languages, mathematics, and science.

Note that quantum physics resolves the ultimate duality between the spiritual and the material in the 'I/not-I' relationship (in religion this is, of course, simply 'God').

Now we might consider there are three categories of knowing: 1) to know what one knows, 2) to know what one doesn't know, and 3) to know what one doesn't know one doesn't know. Category (3) and to some degree (2) is in the expansion cycle of the acquisition of knowledge (whereas (2) mainly requires consolidation, as explained previously when referring to the 'paradox of progress'). One might see that imagination and humility are required to in- voke (3), an apparent nothingness in which we allow thinking and ideas to unfold and develop into (2) and (1). Thus in (1) we have knowledge; in (2) we have knowledge of the existence of areas and subjects that we are not familiar with; and (3) we might be con-sidered to have knowledge that there can be something more. As indicated, this introduces the definition of progress—expansion and consolidation. This was discussed previously, which when im-balanced leads to cessation of progress; the quantitative takes over the qualitative. Thus (3) is an essential state for opening the mind for the expansion factor of progress. Unfortunately today arrogance inhibits this creative state and prevents the imaginative process, bringing about stagnation.

The initial step in considering the acquisition of knowledge is the observation, which can take both the mental as well as the physical approach; that is, subjective or objective, giving rise to the theoretical and practical understanding of a subject. What about skill ability. Is this knowledge? This simply indicates that objective knowledge is potential or inherent but the skill in itself is only ability. In this case, skill knowledge is knowing one can do something—not objective knowledge. It doesn't become knowledge until it is explained, at which time it becomes objective. Subjectively one knows how to make the movement; for example throwing a ball.

Furthermore, models of knowledge and subsequent understanding of knowledge unfortunately are not independent of the acquisition system of the knowledge itself. Suppression of knowledge will inevitably encroach on the structural direction knowledge takes, as exposed by quantum physics: 'What is observed is in the context of whom or what is doing the observing'.

Now religion emphasises the *subjective* and science the *objective* and, for example, the development of knowledge was not free from the church's influence. Until more recent centuries, theology had a powerful control over knowledge, restricting the evaluation of methods of inquiry and suppressing logical approaches. St. Augustine exemplifies the scholarly religious outlook, who stated that the mind is a receptor for divine wisdom and it shares in the glory of God; through it one can acquire knowledge which is unknowable through the bodily senses. Moreover, by medieval times church doctrine dominated university faculties and restricted learning, of which knowledge was pursued through faith.

Nevertheless, the 16th and 17th Centuries witnessed the demonstration of empirical science, and Newton stressed observation and induction; that is, observation and the quantification of observables through experiment. Hardcore materialistic science was in the making, and metaphysics was no longer regarded as a respectable scientific pursuit; knowledge through intuition or religious means was rejected. The subjective mode was no longer acceptable, and empirical objective science became dominant. However, as per the New Science it failed to comprehend the observer/-observed relationship revealed by quantum physics.

In the 19th Century, even psychology was considered dubious and debatable as a valid subject for study, in particular, since a physiological approach to human behaviour had been developed. Fortunately British philosophy provided strong support for the study of psychology and with further developments, in which the various schools of psychology—functionalism* and structuralism,* psychoanalysis, gestalt,* and behaviourism*—influenced by philosophy and empirical science, and evolving towards a methodological system, established a position in science for psychology, although heavily biased towards behaviourism.

> [* Functionalism: mind has a physical basis and the brain is causal.
>
> Structuralism: behavioural effects from language structures.
>
> Gestalt: emphasis on wholeness, organised whole, focus on the whole, not reducible to parts, it is holistic (compare holism) in its therapy applications (New Science compatible).
>
> Behaviourism: materialistic approach, akin to 'reflexes of the brain' and avoids the nature of mind and consciousness.]

In contrast, Eastern philosophies, which began having an increasingly greater impact on the modern Western world, are based on principles that are incongruous with scientific methodology. Hinduism and Buddhism advocate extinction of individuality, of desire and material dependency, and return to universal unity or spirit.[1] The intellect and sensory data are considered fraught with illusions and futile for evaluation of truth.[2] Thus from this outlook, empiricism, a procedure of observation, hypothesis and experimental verification, is unreliable.³ The contrasting functions are that the Western approach acquires information using physical senses, which receive input from the external objective world (inside the context, which means senses have the same order of mechanics as the external world). However, whereas the Eastern approach is a development of consciousness in which information is acquired internally and subjectively but is nevertheless the result of direct contact with the universe; it is experiential and not understood or accommodated by modern science.⁴

In the New Science, we see that the need for science to follow empiricism is based on an assumption which excludes nonlinearity (the internal, inner-space hierarchy). We have seen that there are many space-time frames superimposed and higher levels of unity which require a higher-dimensional logical approach not amenable to either 3D Aristotelian logic or materialistic empiricism. Classical science is based on a 3D model in which space-time is homogeneous, contains particles or mathematical infinitesimally small points of energy, and in which all things are analysable in terms of these separate particles, a procedure referred to as reductionism.

As alluded to earlier, we also may now note that Plato's, 'All knowledge is inborn,' is not so incongruous. In fact as per quantum physics, if we imagine removing all life from the universe (as we have seen), on the basis of the 'I/not-I' relationship and the single origin of mind, the universe would immediately disappear. Objectivity is unconscious subjectivity.

What is not recognised is that in this model each point of energy, which in reality is the quantum (a whole unit of energy, a wavelength), has an internal structure that connects to higher orders through the strata of a holographic fractal system; all space-time levels are connected internally as dealt with in Sections 12 and 13. This is the inner consciousness so discredited as a method of knowing. It gives direct contact with the universe internally (also is the 'within' in the Bible), as opposed to the external-world view which occurs across space using physical senses and is truly not objective but contextual, meaning that what one perceives is governed by the method of detection or perception that exists on the same level of consciousness or degree of order). Owing to this contextualness of all ego-objective observations, sensory data becomes unreliable.

It is thus essential that mental empiricism is (at least eventually) introduced in this new light. It means that man *is* capable of knowing by thought alone. However, this ability is in its early developmental stage in man and it must at our level be combined with deductive and inductive reasoning, along with experimentation when possible and, in particular, a thorough study of all available knowledge in the area of investigation. We can see that the Eastern viewpoint after all is closer to truth than Western materialistic empiricism.

One can see quite logically from this that the very tenets of scientific methodology, which enforce objectivity (separation of the mind), only aggravate the difficulties, causing the poles to become more separate and establishing more unconsciously the opposite pole (the 'not I'), giving greater unconscious subjectivity clearly expressed by the 'I/not-I' relationship. Thus even the objectivity of knowledge is based on quantum physics, the 'I/not-I' relation of observer/observed and it becomes clear that the 'not-I' is a description of our 'external' unconscious mind, individually and collectively. This 'I/not-I' is an infinite fractal all the way back to Source, the 'I'. This means from the start/top of our fractal gradient (compare tree trunk) there is only mind, the quantum realm of infinite possibilities, which has in effect divided up into the observer/observed relationship for the exploration of its own infinite subjectivity. In this process the wholeness of the Absolute is being individualised. Thus regarding scientific methodology we can know that the more science asserts objectivity in the experimental set-up, the less the subjective interconnectedness (direct connection between observer and observed) and the greater limitation is imposed on the results. We have already given examples, such as not detecting the higher-dimensional aspects in the Newtonian measurements caused by quantum reduction (collapse of the wave function). The more the natural subjectivity enters into the observation the more there is communication with the environment as the unconscious collective mind and the more the results will transcend 3D (but clearly this will mean the mode of observation must change).

We see that the different psychologies and philosophies support different aspects of the two categories, subjective and objective, with irresolvable conflicts, however, as already clarified we must introduce the variable of how advanced is the civilisation. The more advanced the more the experiential replaces the objective mode of knowledge (ultimately no objective experiments are necessary)—which is obvious, as the environment/universe is recognised as mind. We might call this observation 'mental empiricism'; more direct, it bypasses the relativeness/illusions of physical empiricism.

Can knowledge be true or not true information? When it is found to be not true it shouldn't be considered as knowledge, since knowledge is ultimately knowingness, which is experiential. When it is subjectively experiential it is equivalent to knowing; it is a direct duplication of the creation (of which this duplication procedure unfortunately is meaningless using the mainstream establishment models of science and philosophy). Historical recordings/memories would be considered knowledge but if they were in the objective category (not having been experienced) they could be wrong, and would not constitute knowledge. If, however, they were subjective and direct memories they would be true. When the information is not true it is merely data about something and does not qualify as knowledge, which must be experientially factual. Furthermore, when proving objective data experimentally, even when dealing with illusions (but unknowingly), measurements can be regarded as knowledge because as we have seen, this can be correct knowledge in the strict relative sense.

There can be considered to be qualities of knowledge. The higher the quality, the higher the degree of order and commensurate relative degree of truth, such as in the fractal system (a branch is a higher order than the twig fractal). For example, knowledge of the application of Newton's laws as opposed to a higher-order/quality or more complete law, which extends and bypasses Newton's laws in more coherent conditions (see Section 19).

Thus qualitative knowledge follows the fractal principles of degrees of order—which is a pyramidal system. The top of the pyramid, say, the tree trunk, or president in the company example, is common to all and a reference for the lower parts. However, on the lower levels it may be difficult to know about the highest level; the frequency is higher and the energy is more expanded. In fact, in the lower viewpoint (of the higher effects), in some cases there will be greater apparent randomness. (It is a mathematical fact that the higher the order the more random it appears on the lower levels.)

The ladder analogy can also be used for showing the basic structuring of knowledge. The regulating factor is progress, which divides into expansion and consolidation.[5] In the ladder analogy we can be rather stringent and place our level of knowledge along the bottom rung of the ladder—partly since we emphasise 3D only.

Consolidation can now be considered to be progress along the first rung. This would be consolidation of pieces of information, observations, etc. It would include formulating principles and laws of science; it would bring organisation, coordination, association, integration, technology, etc. The ego and its need for security encourages this type of progress, and will resist expansion, which would be to the next rung. This is somewhat like fractals and can in fact be entirely due to a fractal change (say, from the 3D spectrum to 4D).

The second rung of the ladder now gives characteristics of knowledge of a higher order, a higher organisation. If we are considering a theory it may now be modified to increase its generalisation to explain more phenomena on the higher level. Note that 'generalisation' is a test of truth in physics (see Section 4). On the second rung we would then concentrate on consolidation at this higher level of knowledge (experience at this higher level would be such that the subjective/objective ratio is greater, and that means the observer is more aware of the environment and is in phase with it more, enabling a greater unity and harmony and natural control over it). This procedure would continue to the top of the ladder, which strictly would be the highest universe/cosmos level of existence prior to the Absolute (see Section 6) and just above would be our single mind before it divides into observer/observed.

Clearly, in acquiring knowledge and in the process of learning, integration of small elements of knowledge occurs. Organisation, association, coordination is involved in this process, always leading to greater wholes. However, since we exist in a universe in which everything obeys the laws of fractals, which we have explained is a system of dimensional degrees of order (similar to the fractal relationship of twigs relative to branches, which in turn are relative to larger branches, etc.), then there are levels of integration, and further potential integrations at each stage (fractal branch, or rung of the ladder). As long as we recognise that systems of observation, in which what one observes is in the context of whom or what is doing the observing, are simply relative (to a reference or contextual level), then progress will occur at an ideal rate.

Another analogy for this is to imagine the letters of the alphabet are units of construction in general. The lowest degree of organisation and intelligence would be that we can only use single letters in communicating, such as 'a', or 'I' (imagine that there are many single letters with meaning; this is only an analogy). As we expand to the next fractal level, we can now use letters, such as: to, of, it, as, etc. New possibilities, concepts are available to us, and so on for further letters.

An example in real life would be the change from a degree of coherence to a higher degree, such as in the Newton's laws example earlier, or for instance, consider integration of white light to create the coherent laser beam, with greater possibilities and applications than just a beam of white light in which the rays are not in phase (Appendix B). Note that the frequency of the coherent beam is in a higher spectrum (fractal level) than the frequencies of the white light. Science doesn't detect this greater coherence (the method of detection will collapse the wave function). Thus higher orders of knowledge are only revealed by a corresponding level (or higher) of observation.

All knowledge is ultimately experiential/subjective/direct/-beingness. We have been deceived into believing that the immediate external world is complete and objectivity, or the separation between observer and observed, is real; subjectivity can then be suppressed and the outcome is a two-lobed brain with an education supporting the left side, intellectual, rational, quantitative, objective and discouraging the right brain: intuition, imagination, etc.[6]

Inevitably there is an accuracy and quality to knowledge. We tend to neglect the order of references that people believe in, regarding the validity of information. Recall that all knowledge and energy are contextual. We have no scientific theory of life and we are taught that life comes from matter or, in other words, life is in the context of matter; yet science recognises that life is a higher order than matter. It makes a significant difference to the ensuing knowledge when comparing the results with the more common belief (from religion, New-Age, metaphysics) that matter is in the context of life—in other words, life comes first. Similarly we place life in the context of death (cycles of life and death is considered a reality), the New Science, of course, reverses these; life is perpetual, the aliveness characteristic is an absolute. A further example is that time is in the context of the Absolute; the Absolute is eternal but this doesn't mean that 'it goes on forever'. There is basically no time. Relatives are in the context of less relatives (twigs to branches, etc.) and ultimately in the context of the Absolute. A single particle in the universe must be defined in the context of its space (environment/-context). Fractals are a contextual system, for example, twigs function within the context of branches, etc. The higher is the order in the fractal system, the greater the truth (contains more information). All relatives must be in the context of an absolute, a true zero from which to make judgements, evaluations and measurements.

As a final comment, the Absolute obviously must be 'contained' in everything similar to the ocean model of quantum physics, in which waves, eddies, wave patterns (all shapes with motion) form all manifestation, are made up of the ocean water.

Thus all knowledge is ultimately subjective. Plato was correct when he stated that 'all knowledge is inborn;' however, he added that 'its apparent acquisition by experience was an illusion.' The inborn aspect is subjective, but the experiential is objective knowledge; a form of representation of the 'not-I', the unconscious objectified, or separated out from the basic 'I'.

Notes

1. This statement is referencing a gradual development over a long period of time. Clearly the ego at our level needs to be transcended from its emphasis on self, to a more expanded state, which encompasses increasing degrees of all creation (compare the fractal tree analogy of twig merging back to the tree trunk, 'ascending' the fractal levels back to Source). Material dependency will obviously be superseded and compulsive emotional desire will sublimate.

2. The intellect (left brain), in the analytical mode, focuses on the importance of the part, and without the assistance of the intuition (right brain) does not recognise higher unity or greater contexts of information (in fact its focus on the part collapses the wave function of the greater truth). However, the latter statement is not recognising that the illusions are relative to the fractal gradient and if 'science' knows this, a natural progression of knowledge will take place along an ascending gradient, reducing relativeness as the context expands, ultimately returning to the Absolute. Thus as long as truth is recognised as relative there is no problem (however, this is the New Science

and current mainstream science is stuck in 3D, as explained in this book).

3. Empiricism is based on the external observation which includes using physical senses that are of the same nature as that which is being observed, that is, inside the context. Hence the quantum physics statement that the observer is part of the experimental set up. But one can step outside, level by level. The twig, say, has no idea that it is connected to the branch (higher knowledge) and that the apparent zero (in any measurement) at the junction with the branch is an illusion and is, in fact, a relative value as it is attached to the branch (and receiving information from 'above').

4. The so-called junk DNA is known to be due to advanced mutations —base pairs have been knocked out.* This has resulted in our carbon body mutation with two brain lobes instead of one, which filter/-format the interfacing mind/consciousness, causing the well-known left and right-brain characteristics separated out, so that our manipulated educational system can reward and develop the left (intellectual, logical) and suppress the right (intuitional). The main purpose being that the isolated left brain can easily be programmed.

* Book, *Engaging the Extraterrestrials* by N. Huntley.

5. Article: *The Paradox of Progress*, in The New Education series. www.nhbeyondduality.org.uk.

6. The human race's genetics has been programmed to evolve a two-lobe brain, separating the intellect (left brain) from the intuition (right brain) then supporting the left brain and suppressing the right since the left brain is much more vulnerable to mind control (see chapter on Stonehenge in the book *Engaging the Extraterrestrials*).

BIBLIOGRAPHY

Bearden, T. E. Solutions to Tesla's Secrets and the Soviet Tesla Weapons. Milibrae, California: Tesla Book Company, 1981.

Bohm, D. Wholeness and the Implicate Order. London: Rootlege & Kegan Paul, 1980.

Capra, F. The Tao of Physics. New York: Bantam Books, 1977.

Chown, M. Quantum Theory Cannot Hurt You. London: Faber & Faber Ltd, 2007.

Cooper, A. P. Our Ultimate Reality. Ultimate Reality Publishing, 2007.

Davies, P. Other Worlds. New York: Simon & Schuster, 1980.

Davies, P. The Cosmic Blueprint. New York: Simon & Schuster, 1980.

Deane, A. Voyagers: Secrets of Amenti, volume 2. Columbus, NC: Granite Publishing, 2002.

Fara, P. Science: A Four Thousand Year History. London: Oxford University Press, 2009.

Feynman, P. R. The Character of Physical Law. London: Penguin Books, 1965.

Goswami, A. The Visionary Window. Illinois: Theosophical Publishing House, 2000.

Gribbens, J. Science: A History. London: Penguin Books, 2002.

Hawking, S. A Brief History of Time. New York: Bantam Books, 1988.

Herbert, N. Quantum Reality: Beyond the New Physics. New York: Anchor Press/Doubleday, 1985.

Huntley, N. The Attainment of Superior Physical Abilities and the New Science of Body Motion. USA: Xlibris Corporation, 2005.

Huntley, N. The Original Great Pyramid and Future Science. London: Author House, 2011.

Huntley, N. A Paradigm for Consciousness (parapsychology dissertation). St. John's University, USA, 1987.

Kumar, M. Quantum: Einstein, Bohr, The Great Debate about the Nature of Reality. London: Icon Books Ltd, 2009.

Norman, E. L. The Infinite Concept of Cosmic Creation. El Cajon, California: Unarius, Science of Life, 1970.

Ouspensky, P. D. The New Model of the Universe. New York: Random House.

Russell, W. The Secret of Light. Virginia: University of Science and Philosophy, 1974.

Rucker, R B. Speculations on the Fourth Dimension. New York: Dover Publications.

Talbot, M. Beyond the Quantum. London: Bantam Books, 1987.

Toben, R. Space, Time and Beyond. Toronto: Irwin Company Ltd., 1975.

Vedral, V. Decoding Reality. Oxford: University Press, 2010.

Wolf, F.A. Parallel Universes. New York: Simon & Schuster, 1990.

Wolf, M. The Catches of Heaven. Pittsburgh, PA: Dorrance Publishing, 1996.

INDEX

A

Absolute, 13, 15, 23, 37-39, 41, 53, 85, 103, 108, 164, 312, 316
action concept, 78, 91, 97, 99
aesthetics, 49, 147, 153, 155, 161
aether, 119-121, 217-224, 230-234
  covariant, 56, 250
  itself moves, 265
aliveness
  characteristic, 55, 108, 119
analogies
  artist paint, 55
  circle/matchsticks, 54
  programmer/computer, 111
  dog/tail, 48
  ice, 39
  ladder, 12, 77, 131
  meter, 43
  ocean, 15, 40, 83
  see-saw, 44
  train, 43
  tree fractals, 14, 17, 32, 45, 57, 113, 159
  spotlights, 222
  TV screen lines, 51
  water/grooves, 91
antigravity, 226, 228, 229, 237, 242-247, 290, 302
antimatter, 32, 232, 236, 291
antiparticle, 31, 32, 226, 232, 236, 289, 291, 297
antivirus, 133, 271
Aristotle, 304
art, 49, 149, 152, 277
artificial, 128
Artificial Intelligence, 302
artist, 151, 159, 277
arts, 149
Aspect, 27
associationism, 305
atom, 66, 84, 228, 230
awareness, 68, 89
  animal, 90
  automatic, 89
  creative, 89

B

Bacon, Francis, 169
Bacon, Roger, 305
barriers
  unnatural, 268
Bearden, Tom, 298
behaviourism, 310
Big Bang, 12, 62, 118
blueprint, 18, 19, 31, 132
  Divine, 133
Bohm, David, 22, 64
Bohr, 129
Bose-Einstein, 224
boundaries, 205, 212
  natural, 268
boxing, 190

brain, 15, 74, 87, 131, 139,
    156-158
    lobes, 318
    interface, 108
    temporary memory, 108
C
catalyst, 137
centripetal, 239, 241
Cezanne, 159
chi, 190
Clauser, 27
collapse
    wave function, 22, 29, 30,
        49, 113, 154, 211, 224
computer bit, 21, 105, 123,
    190
concert pianist, 187
consciousness, 103-109
    relation to Absolute, 108
conspiracy, 160
Copenhagen, 224, 287
Copenhagen Interpretation,
    113, 288
Cosmos, 15, 76, 117, 130
counteraction, 117, 120, 125
covariance, 38, 40, 57, 256
Creation, 12, 19, 32, 54, 73, 81,
    98, 115
cricketer, 186
crop circles, 213
curvature
    space, 39, 88, 237
cybernetic, 74, 89, 103, 138
    interface, 78
    ultimate, 106
cyborging, 160

D
Darwin, 19, 27, 33, 82, 118,
    266
differentiation, 18-22, 273
Dirac, 232
DNA, 31, 49, 83, 88, 129, 318
    junk, 31, 129, 131
    12 strands, 51
doctorate, 20
doughnut, 240, 284
dualism, 15, 38, 39, 86, 130

E
Earth
    tilt, 31, 129
education, 11, 35, 87, 130, 169
Einstein, 15, 27, 28, 43, 54, 56,
    173, 224, 257
    great debate, 129
    general relativity, 39, 81,
        237
electric field, 233, 247, 292
electromagnetic, 24, 34, 177,
    238
electron, 24, 223, 230-235,
    291, 296, 235
    shells, 246, 250
emergent, 153
emergent software, 153, 276,
    300
e-meter, 144
empiricism, 266, 304, 318
    physical, 266, 304, 306, 310,
        311
    mental, 312
Entropy, 63
ESP, 49
ethics, 50, 119, 194

evolution, 31, 32, 131, 192, 299
existence, 299
experimental psychologists, 17, 183

F
Father, 127
feed-back, 91, 186, 188-189
female mind, 127
  magnetism, 127
finite forms, 81
fission, 185
football, 75, 182, 198
force-field, 34, 233, 247, 291, 297, 301
formatting, 76, 81, 82, 87, 108
fractals, 13, 34, 44, 60-70, 72, 117, 151, 179, 193, 206, 261
  hierarchy, 151
  selves, 46
  tree, 14, 17, 57, 69, 113
free energy, 145, 228, 234
free fall, 233, 242, 247, 252-255, 291
free will, 61, 64, 93, 126, 132, 204, 268
functionalism, 310

G
galaxy, 22, 31, 47, 66, 130, 226
gauge invariance, 259
generalisation, 38, 40, 228
genius, 198
geometric, 66
  4D, 230
  intelligence, 211, 212, 228
gestalt, 310

God, 12, 38, 85, 117, 131, 308. 309
golf, 186
golfer, 187, 189
gravitational field, 233, 237
gravity, 39, 217, 224, 229, 233, 237
Grieg, 166

H
harmonic science, 51, 299
healing, 135
heretical, 249, 302
holism, 306
hologram, 55, 60, 66, 75, 132, 267
  interface, 126
holographic, 16, 21, 96, 104, 105
  4D template, 21, 76, 114
  template, 19, 21, 82, 83, 103, 121
  in time, 104
  insert, 268
  terms, 282
  universe, 260
  fractal, 66, 122
Holy, 68, 78, 119, 127
Holy Spirit, 78, 127
hyperspace, 223, 263, 265
hypersphere, 81, 88
hypnosis, 86
hysteresis, 244, 246

I
ideation, 100
illness, 135
impact, 186, 189

inertia, 177, 187, 224, 241, 250
infinite, 13, 38, 56, 61, 62, 64, 69, 119,
Infinite, 81, 86, 125, 141
infinite regression, 23, 38, 118
inner space, 56, 59, 63, 104, 208, 294, 299
I/not-I, 91, 140, 307, 312
integration, 20, 22, 32, 58, 75, 164, 273
intellect, 19, 115, 200, 275
intelligence, 63, 69, 192, 193
Intelligence Quotient, 156, 192
interface, 43, 47, 68, 74, 104, 108

J
Jesus, 132, 275

K
karma, 133-134, 137, 144, 271
karmic, 133, 136, 144
Kelvin, 227
ki, 190
kinaesthetic sense, 21, 97, 100, 105, 114, 115, 176
knowledge, 304

L
ladder analogy, 12, 77, 131, 286, 314
laser, 24, 75, 151, 275, 279
learning
  general, 181, 183, 185
  specific, 179, 181, 183-185, 190, 270
learning patterns, 20, 46, 76, 83, 90, 96, 99, 103-106, 175

Leibniz, 25
Light, 249, 250, 261
  and electricity, 127
light velocity, 36, 56, 250, 257-262
longitudinal waves, 291, 293-295

M
Magdalene, Mary, 132
magnetic, 24, 127
  feminine, 127
magnetic field, 225, 230-231, 263, 292
male mind
  electrical, 127
martial arts, 186, 189
masculinity, 127
mass, 16, 24, 121, 224, 233
  mental, 96
mathematics, 37 43
  base-10, 51
  base-12, 51
Michelson-Morley, 41, 227, 256
Milky Way, 31
mind, 15, 23, 79, 108, 113-115, 141
  can go higher, 33, 78
  control dogmas, 28, 82
  universe, 34, 67
mind computer, 21, 89, 123
mini-black/whiteholes, 217-222, 251, 264, 296
mini-blackholes, 242
Mozart, 206
multiple standard, 50
multiverse, 268, 286
muscular, 168, 175

muscular system, 168, 187
music, 149, 152, 153
   rhythm, 153-156

N
nativism, 306
natural, 12, 18-19, 68, 128-129
Nazis, 228
NET, 272, 268
Newmann, 225
Newton, 39, 43, 177, 224, 227, 242, 248, 309
Newtonian, 43, 49, 224, 239, 242, 253-255
Newtonian force, 291, 292
9/11, 296
Nobel Prize, 298
nodes, 219, 221, 222, 242, 246, 251-255
non-harmonic, 151, 177, 269, 299
nonlinear, 55-57, 114, 210
nonlinear system, 21
non-quantifiables, 54, 125, 150, 175
nuclei, 247, 256, 278

O
observer, 27-29, 33, 34, 123, 140, 258, 262
oscilloscope, 151-152, 174
Ouspensky, 236

P
paradigm, 227, 320
particle and antiparticle, 32, 232, 236
   combine, 31, 227, 291, 297

perfection, 119, 160, 163
perpetual, 61, 87, 217, 220, 296, 301
perpetual motion, 49
Philadelphia experiment, 34
physical mobility, 73, 176
physical training, 175
piano, 96
piano-playing, 273
   technique, 186
Picabia, 159
placebo, 138
Planck, 48
Plato, 304, 306, 317
polarities, 279
   within polarities, 32, 243, 289
Polifka, 225
Pollock, 171
positron, 231, 232
power transfer, 187
practice, 97, 176, 188-190
prana, 190
preferential formats, 37, 59, 211
Priore, 142
psychic, 194, 195
psychologists, 84, 96, 136, 183
psychology, 17, 20, 104, 179, 182
psychotherapies, 143

Q
Quality, 18, 149, 150, 170-173, 197
quantum
   computers, 176, 179, 187-189
   leap, 286

physics, 15, 26, 36, 60, 113, 123, 220
quantum regeneration, 13, 150, 214, 273

R
rationalism, 305
reaction times, 97, 98, 187
realism, 161-165, 171
reality
   artificial, 12, 128-130
   mutated, 128, 131, 291
   natural, 128, 129, 132
reductionism, 305
relative zero, 28, 41, 42, 212
relativity, 28, 43, 219, 227
   general, 15, 39, 81
   special 36, 56, 258, 263
religion, 12, 25, 68, 128, 308
   and science, 12, 130, 328
responsibility, 133, 137, 141
reverence, 194, 206
robot, 34, 84, 92, 160
   analogy, 111, 114
Romans, 132
Russian, 293
   torsion, 293

S
satellites, 251, 253
scalar, 21, 58, 207, 210, 233, 290
   power, 293
   sonic weapons, 292
   coupling, 293
scalar fields, 177, 243, 247, 248, 254
   waves, 58, 177, 199, 207, 291

Schauberger, Viktor, 225, 228
Schrodinger, 224, 250
Science and education, 11, 87, 110
scientific method, 15, 25, 28, 45, 292, 312
Searl, 225
self-reference, 48-49
self-referencing systems, 42, 86
Severini, 167, 168
skills, 83, 170, 186, 191, 210
   impact, 186
snooker, 180, 182, 191
soul, 46, 59, 77, 87, 111, 168, 171, 172
   over-soul, 88, 173
spacecraft, 237, 253, 278
specificity law, 183
spirit, 78, 91, 125, 310
spiritual, 85, 129, 170, 307
sprinter, 181, 190
stability, 25, 72, 117, 125, 126, 225, 270
   instability relation, 117, 125, 126, 270
standing wave, 21, 57, 222, 293
stealth, 34
structuralism, 310
subjective, 202, 209, 305, 307
   objective relation, 25, 61, 113, 117, 209
subjectivity, 12, 15, 34, 69, 308
Sun, 246, 293
superimposed, 56, 57, 63, 68, 72, 152, 178
superposed, 56

supersymmetry, 228, 283
symmetry, 37, 38, 232
symptoms, 137

T
Taimni, I.K., 53
talents, 194, 199
template, 141
Tesla, 292, 293, 297, 302, 319
Theory of One, 73
Toben, Bob, 36
toroid, 228, 230, 240
training, 175, 176-182
triad, 61, 102, 121, 123, 220
Trinity, 61, 123, 127

U
UFOs, 31, 118
Unified Field, 55, 78, 84, 120, 127, 206
unity, 18, 22, 24, 31-32, 74, 94, 118, 156
   of particle and antiparticle, 227
   over-unity, 225
Unmanifest, 53, 118
Urantia Book, 61

V
Van Gogh, 161
vector, 218, 219
velocity of light, 36, 48, 56, 250
virtual particles, 19, 218, 222, 251, 264, 290
vitalism, 306
vortex, 31, 66, 130, 217, 223-228, 235, 239, 241
   4D, 226, 228, 232, 244, 283

counter rotating, 225
dual, 31, 130, 225
secondary, 230, 245, 284
vortex model, 225, 227, 250, 260
visualising, 283
vortices
   higher dimensional, 287

W
wave function
   collapse, 22, 29, 30, 49, 113, 131, 154, 188, 211, 224, 287, 313
wave/particle
   dilemma, 26
waves
   transverse, 238, 290-297
   longitudinal, 58, 238, 291-297
weight lifting, 181
Wheeler, John, 33, 218

THE EMERGING NEW SCIENCE
volume 1

CONTENTS

PART ONE
THE NEW SCIENCE
1. Introduction to the New Science
2. Is Science Really Progressing?
3. Science Problems
4. Fundamental Difference between Orthodox Science and the New Science
5. A Glimpse into the New Science
6. Features of the New Science
7. Orthodox Science and its Limitations
8. Interconnectivity of the Multiverse
9. Quantum Physics
11. Harmonic and Non-Harmonic Sciences
12. The Mind, Chakra System and DNA

PART TWO
THE FRACTAL TREE
13. Science and Religion
14. The Evolution of the Species
15. Science and Technology

16. The Infinite Energy Source
17. Ego/God Dichotomy
18. Consciousness and the Paranormal
19. Ascension

PART THREE
EDUCATIONAL FALSEHOODS
20. There is No Life After the Death of the Physical Body
21. Scientific Methodology is the Only Acceptable Method of Acquiring Truth
22. The Origin of the Universe: Big Bang Theory
23. The Origin of the Species: Theory of Evolution
24. The Mind is a By-product of the Brain

PART FOUR
THE NATURE OF CREATION
25. The Nature of the Universe, Life and Evolution, and their Purpose
26. Dualism and the Anthropomorphic Principle
27. Creation From the More Scientific Viewpoint
28. Computer Systems
29. Harmonic and Non-Harmonic Technologies

PART FIVE
THE PARADOX OF SOMETHING AND NOTHING
30. First Cause and Infinite Regression
31. Bringing in Quantum Computing

PART SIX
QUANTUM REALM
32. Introduction
33. Quantum Reduction and Quantum Regeneration
34. Resolution of the Great Einstein-Bohr Debate
35. Secondary Quantum Reductions

PART SEVEN
COLLAPSE OF THE WAVE FUNCTION
36. Learning-Pattern Application
37. Creating (Selecting) One's Own Reality?
38. Conclusions

PART EIGHT
FURTHER RELATED TOPICS
39. Quantum Teleportation
40. Free Will
41. Geometric Intelligence

APPENDICES
42. APPENDIX A: The Solar System
43. APPENDIX B: Conflicting Creations
44. APPENDIX C: The Subjective/Objective Illusion
45. APPENDIX D: Existence and Evolution, Quantum Reduction and Regeneration
46. APPENDIX E: Overcoming the Velocity-of-Light Limitation

BIBLIOGRAPHY

INDEX

www.ingramcontent.com/pod-product-compliance
Lightning Source LLC
Chambersburg PA
CBHW021349210526
45463CB00001B/43